An Introduction to Structural Analysis

The Network Approach to Social Research

S.D. Berkowitz

Department of Sociology
University of Vermont

Butterworths
Toronto

An Introduction to Structural Analysis: The Network Approach to Social Research
© 1982—Butterworth & Co. (Canada) Ltd.

Printed and bound in Canada
5 4 3 2 1   2 3 4 5 6 7 8 9/8

Canadian Cataloguing in Publication Data

Berkowitz, S. D. (Stephen David), 1943–
    An introduction to structural analysis

Bibliography: p.
Includes index.
ISBN 0-409-81362-1

1. Social structure.   2. Social groups.
3. Sociological research.   I. Title.

HM131.B47   301   C82-094241-3

The Butterworth Group of Companies

Canada:
Butterworth & Co. (Canada) Ltd., Toronto and Vancouver

United Kingdom:
Butterworth & Co. (Publishers) Ltd., London

Australia:
Butterworths Pty. Ltd., Sydney

New Zealand:
Butterworths of New Zealand Ltd., Wellington

South Africa:
Butterworth & Co. (South Africa) Ltd., Durban

United States:
Butterworth (Publishers) Inc., Boston
Butterworth (Legal Publishers) Inc., Seattle
Mason Publishing Company, St. Paul

To My Grandparents:

Adolph Berkowitz	*Edward Coplan*
1880–1960	*1891–1981*
Pauline K. Berkowitz	*Hannah R. Coplan*
1891–	*1890–1961*

Contents

4: Community-Elite Networks and Markets: Structural Models of Large-scale Processes 125

5: The Future of Structural Analysis: Some Conclusions 149

Preface

This book is intended as a brief and relatively nontechnical introduction to the key concepts, central intellectual themes, and principal methodological techniques of a new approach to social science: *structural analysis* or, as it is sometimes called, *network analysis*. Its purpose, as a result, is quite different from the purposes of other texts which focus on the content or subject matter of a particular discipline.

The term *structural analysis* does not refer to a closely defined body of "facts" which, once mastered, will prepare the reader to apply his or her newly acquired knowledge in operating a day-care center, understanding other people's erratic behavior, or speculating on the stock market. In this sense, structural analysis is not like early childhood education, psychoanalysis, or commerce and finance. Moreover, while structural analysts have had things to say about corporate systems, tribal kinship, urban environments, patterns of scientific training, and community structure (among other things), structural analysis is really not concerned with these problems in themselves, as economics, anthropology, urban studies, and sociology are. Although structuralists often employ mathematical reasoning and formal models in their work, structural analysis is not even best thought of as a branch of applied mathematics or logic. In fact, unlike most traditional academic areas, structural analysis cannot be adequately defined by either its subject matter or its tools—alone or in combination with one another. Structural analysis, in this sense, is neither a conventional field nor even a subspecialty of the kind usually found in a traditional university curriculum.

What, then, is it exactly? And, more to the point, why should someone want to be "introduced" to it? The first of these questions is relatively easy to answer in a general way: structural analysis is an approach to theorizing about, representing, and analyzing social processes which emphasizes their systemic character. It is, in other words, a transdisciplinary paradigm for doing research. As such, its practitioners may be housed in a variety of different fields, interested in a number of different problems, and even employ a number of different kinds of models. Most structuralists, for historical reasons, refer to themselves as sociologists, but this is more a matter of convenience than anything else: in principle, almost any social science problem can be treated in a distinctly "structural" way.

The second question is even easier to answer: we live in a world which is increasingly complex and interrelated. Most of the tools or approaches commonly used in conventional social sciences were not designed to deal with interdependent phenomena. Indeed, they assume exactly the opposite. When conventional sociologists select a sample, psychologists administer a test, or economists model the demand for a good or service, they normally assume that the people selected, the subjects tested, or the consumer preferences represented are independent. This is reflected in their choice of problems, selection of models, and preferred ways of interpreting their results. Whenever "normal" social science of this kind encounters problem areas where this assumption of independence is not met—that is to say, whenever "reality" does not lend itself to being described or modeled in this fashion—these conventional approaches begin to break down. Structural analysis, by contrast, was specifically developed in order to examine precisely those highly organized, interrelated, and persistent or patterned forms of social activity with which conventional approaches and methods are least prepared to grapple.

The larger objective of this book, then, is not simply to introduce the reader to structural analysis, but to invite him or her to start thinking structurally. It begins by describing the origins of this new paradigm in a set of historic problems in sociology, economics, anthropology, and general systems theory which cannot be adequately dealt with by traditional approaches or models. It then outlines how graphic devices, called *networks*, may be used to resolve or at least to clarify these issues. This discussion is followed by a brief focused overview of the broad objectives of scientific inquiry and the role of structural analysis within them.

The chapters which follow examine a series of substantive areas in which the application of structuralist concepts and methods has yielded new and sometimes strikingly different results: studies of kinship and friendship, markets and the economy, communities, stratification, and large-scale social processes. Each of these is intended to acquaint the reader with important theoretical and methodological dimensions of structural analysis, as well as its specific substantive contributions to social inquiry. Finally, the last chapter attempts to draw these strands together and to forecast the directions which social scientists doing structural research are likely to follow in the future.

I take pleasure in thanking the people at Butterworths—Peter Horowitz, David Hogg, and Anne Butler—for their timely support and encouragement. I am indebted to Anatol Rapoport, Barry Wellman, and Harrison White for careful and critical reading of the first four chapters of this book, and to Ron Breiger, Peter Carrington, Lin Freeman, Greg Heil, Nancy Howell, Edward Laumann, Stan Lieberson, Jack Richardson, Lorne Tepperman, and René Thom for many useful suggestions on parts of the early drafts. Joel Levine, whose typically insightful comments helped eliminate a lot of fuss and feathers from one part of it, and his students at Dartmouth College who acted as unwitting guinea pigs for the penultimate draft, deserve my profound thanks and sincere sympathy, respectively. H. Gilman McCann's sage advice and careful help in proofreading facilitated the final stages in the preparation of this book, for which I am profoundly grateful. My students—in particular, Ross Baker, June Corman, Greg Knapp, Edward Lee, Rod Nelson, and Catherine Rogers—added to the clarity of the present account through timely comments on embryonic versions of the arguments and concepts which are presented in detail here. The Structural Analysis Programme, which provided financial aid for typing and xeroxing much of the manuscript, and my colleagues in the Department of Sociology, University of Vermont, who provided me with the social support needed to complete it, richly deserve my thanks. Shawn Berkowitz and Margot Bowlby's careful work on the index was greatly appreciated. As always, without my family's tolerance of irregular hours and father-absent weekends this book would not have come to fruition.

1

What Is Structural Analysis?

BACKGROUND

The idea that social systems may be *structured* in various ways is not new.[1] In fact, all of the established social sciences have evolved some notion of *structure*. But, until recently, no field had taken this idea of a *regular, persistent pattern in the behavior of the elementary parts of a social system* and used it as a central or focal concept for understanding social life. Indeed, even today, the notion of an overarching system of relationships among the parts of a social system is quite marginal to most conventional social science frameworks or *paradigms*.[2]

The reason for this is quite simple: until recently, no one had a good way of representing, let alone measuring, structure. This is critical because scientists are trained to distrust things that cannot be either observed and measured or associated with something that can. Consequently, while some extremely useful scientific ideas have been kept deliberately vague ("personality" and "climate," for instance), science as a whole strives toward *precision* in definition, careful *specification* of observed phenomena (*operationalization*), and *consistent* measurement. Therefore, until there was a good, practical way of measuring patterned activity in social systems, the notion of structure tended to be overlooked in favor of simpler concepts and means of interpreting observable "reality." When social scientists used the term at all, it was almost always in its most general or global sense.[3]

The tool that changed all of this was a simple graphic device called a *network*. In the branch of mathematics called *graph theory*, one can specify the systematic relationships between and among the elements of a graph ("nodes" or "points") and the lines joining them ("edges" or "relations") and then analyze properties of the resulting patterns.[4] In the 1950s and the early 1960s, social scientists borrowed this idea from graph theory and began using it as a means of describing the relationships between and among people, family groups, corporations, government offices, and a range of other *elements* or elementary parts of complex social systems.[5]

At first, networks (or, more properly, *networks of effects*) were employed as little more than metaphors for the things social scientists were really trying to deal with: a friendship group was *like* a "star" with one central point; a work group was *like* a small "pyramid"; or the spread of a rumor was *like* a "chain." While this line of development eventually led social scientists to examine a number of interesting formal or mathematical properties of networks, it did not add a great deal to our substantive knowledge of how social systems work. In contemporary terms, these studies were simply examining the *morphology*, or abstract form, of the systems they were modeling rather than either the interactions among their elements—what we call *behavior*—or the ways in which morphology shapes and constrains behavior, that is, structure.[6]

This use of networks-as-metaphor, however, had the effect of encouraging researchers to think of social systems and social processes in terms of the *relations* between and among their constituent parts. Earlier paradigms had not done this. The overwhelming majority of social scientists, while they used concepts such as "input," "output," "feedback," "system" and "boundary," which had been borrowed from cybernetics and information theory and applied to sociocultural systems, continued to conceptualize the world in terms of *categories* of independent effects and events. With the advent of network-based models, however, they were forced for the first time to begin to take the idea of a *system* which transcends the individual properties of its elements seriously and to consider *systemic* attributes as real phenomena in their own right.

With this change in focus, it became possible to develop more rigorous and powerful techniques for analyzing social network processes. The first genuinely structural applications of these were in epidemiology. Earlier mathematical models of "contagion"—that is, the process or processes through which disease or disease-like information (e.g., rumors) is spread through a population—were based on the simple logistical notions (a) that the spread of "disease," its rate of increase, is proportional to the size of the "affected" portion of the population and the size of the "unaffected" portion; (b) that all contacts between "affecteds" and "unaffecteds" are equally probable; and (c) that increments in the size of the "affected" population arise only from contacts between "affecteds" and "unaffecteds."[7] Given these assumptions, however, it is mathematically necessary also to assume that an individual, once affected, remains affected indefinitely and that he or she continues to be as infective during all stages of the process.[8] These assumptions are unrealistic in that in actual epidemics, some individuals die or simply recover from a disease and thus are no longer able to infect others.[9]

Beginning in the early 1950s, and drawing upon earlier models of transmission processes in nervous systems, Anatol Rapoport and others started to investigate contagion models in which (a) individuals have a fixed, finite number of transmission possibilities, on average, and (b) once they have transmitted a disease or disease-like form of information, it is possible for them to *recover* and no longer be "infective."[10]

Studies of what Rapoport called *ignition phenomena* of this kind enabled researchers to begin exploring the impact of specific social constraints on the characteristic patterns of information transmission or contact in a population.[11] This work on *biases* in social networks formed the core of the modern approach to interpreting structure in social systems.[12]

While Rapoport and his colleagues were beginning to construct formal mathematical models of network-based processes, a group of anthropologists started to use networks as a tool for simply *mapping* patterns of ties among individuals that fell outside the boundaries of the kinds of organized subgroups that sociologists and anthropologists had traditionally identified within populations (e.g., classes, families, tribes). While, initially at least, these studies were largely descriptive and, as such, did not make much use of the elaborate mathematical tools developed by those interested in contagion, they dealt with issues which were closer to the core traditions in sociology, anthropology, and social psychology than those which were being formalized at the time. As a result, their impact on the theoretical vocabulary and mode of conceptualizing problems in what later became known as *structural analysis* was commensurately greater.

During the early to mid-1960s, these two research traditions—the first, more formal and analytic, and the second, more descriptive and phenomenological—came together in powerful ways. The result was the formulation of what Barry Wellman refers to as the *social network concept:* the idea that social structure is best understood in terms of a dynamic interplay between the *relations* between and among persons (or institutions, etc.) on the one hand, and the *positions* and *roles* they occupy within a social system, on the other. Reversing the conventionally accepted logic of inquiry in social science, structuralists came to argue that social categories (e.g., classes, strata) and bounded groups (e.g., biochemists) could best be discovered by examining the relations between and among social actors or institutions.[13] Rather than beginning with an a priori classification of the observable world into a discrete set of categories, then, they postulated the opposite: begin with a set of relations and from them derive a typology and map of the structure of groups.

The potential impact of this apparently simple reversal in the logic of discovery was enormous. We may draw a distinction between what

Abraham Kaplan calls *pictorial realist* and *instrumentalist* definitions of things. Pictorial realists believe that a rude empirical "reality" exists, and that it only remains for us, like the early explorers, to "discover" and chart it. This is in marked contrast to the *nominalist* doctrine that universal or general forms are products of "mind"—that is, devised by us for interpreting or understanding the world—and are thus "things so named." In social terms, both pictorial realism and nominalism tend to reflect *received* categories and conventional wisdom. A "chair," for instance, is an object which belongs to the set of all objects which (pictorial realism) we identify with our "discovered" model of a "chair," or which (nominalism) we refer to as "chairs" in social practice and language. While pure nominalism is rare in science, requiring, as it does, no correspondence between models and observed reality, pictorial realism is not uncommon.

Kaplan contrasts pictorial realism with a quasi-realist notion he calls instrumentalism which holds that theories and models must be "invented" or "constructed" reflecting facets or aspects of observed reality. It rests on what might be called "the principle of practical measurement": a theory or model is a device which may be *used* in a particular way to interpret reality. An object is, therefore, what it does. A "chair," for instance, is something which may be used for the purpose *we intend* in defining a chair in a particular way (e.g., it may be used to sit on).[14]

By defining categories of actors and bounded groups on the basis of their relatedness, structuralists were, in effect, staking out a position as radical instrumentalists. If, they contended, social science is (or ought to be) primarily concerned with social processes, then the implicit *logic* used in defining objects for study ought to be processual as well. In practical terms, this suggested that the measures employed in defining categories ought to be derived from an analysis of social process, rather than simply received or assumed.[15]

Until there was some way of practically interpreting the social network concept in terms of a method or series of procedures, however, it remained little more than a sensitizing device or what we sometimes call a *heuristic*.[16] Two breakthroughs then occurred at almost the same time: the development of *algebraic* interpretations of social networks and the invention of a mapping technique called *multidimensional scaling* (or MDS).[17]

In 1966 John Paul Boyd wrote a dissertation, "The Algebra of Kinship," which formalized and extended earlier applications of the mathematics or logic of *group theory* to the description of kinship systems.[18] This thesis incorporated ideas first put forward by Claude Lévi-Strauss—the French structural anthropologist—and a collaborator of his, André Weil, in 1949, and by Harrison White in the early 1960s.

White's principal contribution had been to show how an elementary *typology* of kinship forms may be derived from a set of prescriptive rules specifying classes of persons who "ought" to marry, and a set of axioms for defining the problem mathematically. He also pointed to broad potential applications of algebraic techniques in the study of areas other than kinship systems.[19] Boyd extended and enhanced this work by demonstrating how the concept of a *homomorphism*—or a many-to-one mapping—could be used to compare structures. Since, under specific circumstances, the generalized relations among parts of social systems can be shown to correspond to what is implied by the mathematical concept of a "group," the stage was set for a general exploration of the algebra of social structure.[20] His thesis, and a subsequent article on the subject, also brought together a number of similar strands of development in linguistics and in applications of graph theory in small-group research.[21]

Harrison White and François Lorrain then took a number of the ideas which were embryonic in Boyd's work and fashioned them into the first really usable tool for investigating concrete (as opposed to general) relations among the elementary parts of a social network: a technique for uncovering structurally equivalent nodes by homomorphically reducing the relations among concrete sets of social actors. In a landmark paper which appeared in 1971, "Structural Equivalence of Individuals in Social Networks," Lorrain and White considered the *semigroups* formed by the graphs defining specific binary (0 or 1) relations among the members of a concrete, bounded population, and all possible *composition* operations defined on these graphs, that is, those combining algebraically or *compounding* different types of relations.[22] By simultaneously reducing both the objects (nodes) and morphisms (generalized compound relations) of a concrete datagraph into a simplified model of itself, they were able to specify nodes in the original graph which had the same pattern of ties in its reduced representation. Since, by definition, these nodes played the same *role* in the model graph, they could then be said to be *structurally equivalent*.[23]

Lorrain and White's method was able to realize, for the first time, all of the power implicit in the social network concept. First, it operated simultaneously on both nodes and relations. This enabled researchers to explore the part played by a given node (e.g., an individual) both in terms of its *position* within a network and as a point of confluence for a set of *role relations* among nodes. Second, it enabled researchers to deal with a given network at all levels of abstraction, that is, as a set of simple binary relations, as a concrete manifestation of a set of abstract roles, and as part of a mathematically defined entity (*category*). In practical terms, this meant that the relationships between these concepts, as well as among the nodes themselves, could be examined at each of these levels.

Finally, the Lorrain-White method firmly established a framework for considering a range of mathematically possible *orders* of connectedness among parts of social systems which, while not obvious to participants, have direct and theoretically interpretable implications for their behavior.[24]

In parallel with these developments in the algebraic modeling of social network structure, another group of researchers were examining ways in which scaling techniques which do not require prior determination of the "distances" between objects in some pre-measured space (quasi-metric scaling) could be used to represent and interpret networks of ties or flows. Building on earlier work by researchers such as R. N. Shepard and Clyde H. Coombs, other scholars, notably Edward Laumann and Louis Guttman, and Joel Levine, published papers in the late 1960s and early 1970s which applied variations of these techniques to representing the "distances" between occupations, between and among banks and industrial corporations, and to other problems of network flow or structure.[25]

The principal strength of these methods lies in their use of behaviorally determined measures of the relative "closeness" of elements to one another, rather than some arbitrary a priori scaling of distances. This, once again, is consistent with the radical instrumentalist notion that properties of systems should be measured using techniques which do not rest on "received" categories, but, as much as possible, derive from observation of the processes through which social systems create or structure themselves over time. During the 1970s, Laumann and a number of his collaborators—Joseph Galaskiewicz, Peter Marsden, and Franz Pappi—extensively employed scaling techniques of this kind to differentiate the structural properties of community elites in two cities: "Altneustadt" (West Germany) and "Towertown" (U.S.).[26] Their studies not only added to our substantive knowledge of the patterns of integration within communities, but, as we shall see when we discuss them in detail later on, definitively established these scaling methods as a widely applicable and flexible tool for testing alternative hypotheses about the organization of community elites. Levine's use of similar methods to represent patterns of *directorship interlocking* (where a director sits on the board of two or more companies) also gave rise to a number of closely related studies.[27]

By the mid-1970s, then, structural analysis had established itself as a distinct paradigm with at least two alternative sets of tools, a rudimentary theoretical vocabulary, and a number of animating ideas or scientific goals.[28] While a number of important contributions to the development of structural analytic methods had been made by researchers working in relative isolation from one another,[29] two research centers predominated in these early stages of the development of structuralism in North America: first, the University of Michigan, and especially its Mental Health Research Institute, and then, after Harrison White moved there from Chicago, Harvard. By the early to mid-1970s, this pattern had changed, and important foci had appeared at Columbia, Chicago, Indiana, Minnesota, the State University of New York (SUNY) at Stonybrook and Albany, Toronto, and, most recently, the University of California at Irvine and Santa Barbara.[30] Productive research centers had also emerged in the United Kingdom, Germany, and Holland.[31]

CENTRAL ORIENTATIONS

In its present form, structural analysis is far from a homogeneous paradigm. This results, in part, from its diverse origins in areas such as small-group research, general systems theory, epidemiology, "structural" anthropology, interorganizational studies, and political economy. Its diversity is also a consequence of the relatively rapid development of the area and its highly integrative focus: since structural analysis is an approach to the specification of problems, rather than a set of answers or solutions, it can accommodate an unusually diverse set of hypotheses, models, orientations, and styles of reasoning.[32]

Despite this diversity, however, structural analysts share a core set of assumptions and, in their most general form, a common set of animating concerns and themes.

First, structuralists accept the view of science as fundamentally "pattern recognizing": that its object is to create *models* which act as *analogs* for complex physical, natural, or social phenomena which can then be observed in the "real world." Ideally, these analogs should be *isomorphic* or *homomorphic* to the phenomena they are seeking to capture.[33]

A model is a simplified representation of "reality" which has been reduced in both scale and detail. Very simple, small-scale phenomena can, as a result, sometimes be depicted by isomorphisms—or one-to-one representations—of themselves. In most cases, however, some loss of detail in this process of representation is inevitable and desirable. Here, essentially, we devise homomorphisms—or many-to-one reductions— in which non-essential differences between cases or instances of a given phenomenon are suppressed.

Structuralists contend that this process of the creation of analogs for real-world phenomena is fundamentally structural, since its goal is to uncover the essential *form* of observable events or processes.[34]

Second, from this idea that the pre-eminent scientific activity is "model building," it follows that the long-term goal of science is to discover relatively simple, lawlike relationships between observed events. We therefore must apply three tests to scientific statements: *consistency*, *parsimony*, and *falsifiability*.[35]

The principle of consistency has to do with whether or not a given statement is logically coherent: does it "hang together"? Is it, instead, self-contradictory? Does it define its terms unambiguously? Does it clearly state the direction of cause it seeks to imply? Is there a clear chain of reasoning between cause and effect? The test of parsimony rests on the idea that the more complicated an explanatory schema, the greater the probability that it will contain "error." Hence, it holds that scientists should strive for simplicity in explanation. Given a choice between two models which fit a set of events equally well, one should therefore choose the simpler. Finally, scientific statements should be falsifiable, that is, capable, in principle, of being disproven. Not all logically consistent and parsimonious statements can be falsified. Only empirically verifiable statements—ones which can be associated with some clear set of "observables"—are capable of empirical disproof. Thus, the principle of falsifiability demands that, both logically and empirically, it should be possible to accept evidence which runs counter to one's hypotheses. Other statements, while they may be real and palpable in their own way, are not scientific in this sense.[36]

Third, structural analysts accept the distinction between (a) modes of reasoning which assume that the relationships between and among the elements of a social system act to set limits on or constrain the behavior of these elements and (b) ones which hold that individual or elementary forces simply "come together" to form larger systems. We refer to the first as *sociologistic* and to the second as *psychologistic* reasoning. Structural analysis is clearly sociologistic in both its intent and methods.

The simplest way of thinking about the differences between these two approaches is in terms of "wholes" and "parts." Those who employ sociologistic reasoning draw inferences about the behavior of elements ("parts") from aspects of the structure of systems ("wholes"). By contrast, from a psychologistic perspective, systems ("wholes") are nothing more than the sum of the attributes or characteristics of their elements ("parts"). The key, then, to distinguishing between the two is the *direction of inference* each implies. If a model suggests that we examine the structure of systems in order to discover the source of the behavior of its elements, it is sociologistic in intent. If it implies the opposite, it is psychologistic.[37]

This distinction is applicable at any level of organization within a system, and is not restricted to the relative properties of "individuals," and "groups" or "societies." Thus, structural analysts have implicitly recast this classic "problem of levels" in terms of the *direction* of cause between phenomena, and have sought to discover, empirically, what and where, exactly, these boundaries are.[38] Which units may be treated as "elements," in a given case, and which combinations of these may be part of "higher order structures" are thus left as problematic.[39]

Fourth, given its implicit reformulation of a notion of levels, structural analysis also rejects the historic dichotomy between *atomistic* and *holistic* strategies of explanation.

Karl Popper and other philosophers of science who were responsible for this distinction assumed that a system or problem could be treated either *analytically*—broken down into its components—or *synthetically*—composed or combined into a larger whole. Analytic strategies, by their very nature, were seen as atomistic and thus tending to represent systems in terms of the characteristics of and interactions among elements. Holistic explanations, by contrast, were thought of as synthetic and thus tending to focus on "wholes" to the exclusion of "parts."

Structural analysts, however, lay great stress on the *unitary* nature of social processes and, hence, mutual interaction and interdependence of systems, subsystems, and elements. This perspective was well represented in the literature in general systems theory, which is one of the antecedent traditions structural analysis draws upon heavily.[40] However, it was never concretized in a set of techniques or methods which intrinsically demanded that researchers pay attention to this mutual interdependence of "parts" and "wholes." As we shall see later, one of the methods most often used by structuralists today, *blockmodeling*, is essentially based upon an examination of this issue.[41]

Finally, structural analysis, given its transdisciplinary character, easily accepts relevant concepts, ideas, models, problems, or objects for study, and formal techniques from a range of other fields. Frequently, as a result, structural arguments attempt to relate modes of conceptualizing a problem in one conventionally understood discipline to those in another.

Historically, academic specialties were built around classes of events. The perceived boundaries between these events were then thought of as delineating "fields of study." Thus, for instance, economics would be seen, conventionally, as that discipline which deals with economic facts.

The difficulty with dividing up the scientific world in this way, of course, is that there are typically a number of different *dimensions* which we may observe at work in any real-life situation. Hence, while the borders between, say, "German literature" and "English literature" may be relatively easy to patrol, those between sociology and economics would be considerably more difficult to enforce. Any reasonably interesting problem in one social science ought to contain at least some elements of interest to the others.

Structural analysis, because of its emphasis on model building, has played havoc with these definitions of academic "turf." Until recently, the only effective enforcement mechanism for boundaries among social sciences was that each field utilized a different conceptual vocabulary and, for the most part, different models. Thus, to use economists' terms, the transaction costs across "boundaries" were high, since each crossing involved acquiring a new language and set of skills. By insisting on unitary descriptions of social process, however, structural analysts have, to a greater or lesser extent, forced conventional disciplines to come to grips with each other. Moreover, by incorporating explanations and models that draw upon more than one such field, structuralists have exposed points of convergence among the traditional disciplines.

COMMON THEMES

Apart from these broad orientations to scientific work—which they share, in some measure at least, with some of their more conventional brethren—structural analysts are usually also interested in a range of theoretical and methodological issues which are not part of the normal social scientist's repertoire. Many of these concerns, in fact, would fit more comfortably into a curriculum in physics, chemistry, or zoology than one in sociology or economics: "boundaries" within systems, the construction of endogenous units of analysis, "deterministic" versus "probabilistic" reasoning, "static" versus "kinetic" modeling, and so on.[42]

While older social science paradigms sometimes deal with these questions in passing, structural analysts have been forced to tackle them head-on because of the precise forms of measurement normally used in their research. This is extremely important because fundamental epistemological issues that cannot be adequately understood using conventional approaches and tools have been cropping up in various social sciences since the 1850s. Some of these now constitute core theoretical issues in structural analysis.

Two social thinkers—Claude Henri Comte de St.-Simon and Karl Marx—first explicitly argued that there is an overarching systemic relationship between people which transcends their individual or interpersonal relations with one another. Earlier writers had viewed "societies" (or "commonwealths" or "nations") psychologistically, that is, as mechanisms for reconciling the disparate interests or enforcing the rights of individual social actors. The "sovereign" or "state," they argued, was simply a device for lending authority or giving legal force to overlapping sets of individual "covenants between men." While extensions and implications of this so-called social contract theory, both in social theory and the philosophy of law, are extremely complex, they were and are indissolubly linked to the notion of an "individual" as the fundamental and irreducible social unit.[43]

St.-Simon and Marx, however, identified *groups* (or, more properly, "sets") of individuals sharing common social locations and life chances, which acted *collectively* to promote their interests and worldviews within the social order.[44] Whether one focuses on "social classes," as St.-Simon and Marx did, or on other kinds of "social groups" as fundamental units (e.g., "intermediary groups," "status groups"), the epistemological conclusion is the same: there is a *level* of organization within societies which cannot be adequately understood by simply observing *individual* behavior.[45]

Problems of Levels and Units of Analysis

Emile Durkheim, one of the fathers of modern sociology, extended this idea in a famous monograph, *The Rules of Sociological Method* (1895), in which he defined a set of "social facts" which are, simultaneously, "external" to the consciousness of individuals and act to "constrain" their behavior.[46]

During the next half-century, this distinction between "individual" and "collective" action gave rise to two fundamental and interrelated theoretical issues which surfaced in many different contexts: the *problem of levels* and the question of *units of analysis*.

In its basic form, the problem of levels has to do with whether or not Durkheim's theoretical distinction between "individual" and "social" phenomena can be sustained empirically. Is there a body of "facts," corresponding to a set of observable events, which is inherently social? Moreover, if there is, how can we discriminate between social events and nonsocial or individual events?

In practice, this problem immediately suggests another: assuming that we know which *events* are social and which are not, what are the appropriate *units* that one ought to use to capture phenomena at this level? Put another way, how does one go about *constructing* models of "reality" which reflect this distinction?

Conventional social science "solves" both of these problems by fiat: "events" are simply arbitrarily classified as either "social" or "individual," and units of analysis are chosen which are *assumed* to reflect this distinction operationally.[47]

Economists, for instance, have been interested in the way in which individuals, families, firms, and other "decision making" units allocate income among alternative uses (e.g., transportation, education or training, savings). However, empirical studies by economists have shown that individual preferences are poor predictors of the consumption patterns of larger units, for example, "families." This suggests, in turn, that "family preferences" are not simply the aggregate of the individual preferences of members of families.

Following Durkheim's usage, we can say that these economists discovered that *family income allocation* is a "social fact." How, then, is it possible to measure it in some rigorous fashion? The economists' solution is to devise a new unit of analysis, the household, which is presumed to correspond to the effective decision-making unit whose behavior they are trying to represent.[48]

In practice, this presumption is never tested independently. A given definition of a household is simply devised, data are gathered using it, and these data are then aggregated as though one were dealing with individual preferences. The difficulty with this, of course, is that the world does not fall neatly into separate "boxes," and there is no guarantee that any particular definition will reflect the *process* of income allocation which economists had in mind in the first place.

By contrast, the structural approach to these problems begins by devising a measure or series of measures of the *process* one is trying to capture in one's "units." Thus, in the present case, a structuralist would explore alternative types of ties among family members (kinship? dependency? authority?) in order to discover which one, alone or in combination with others, most consistently and coherently reflected the *dimension*—that is, the control over decision making—most salient to the allocation of family income. These measures themselves—or, more properly, the *patterns* resulting from them—could then be scrutinized in terms of their consistency, parsimony, falsifiability, robustness, and so forth. If a given measure or combination of measures met these criteria, it would then be used to specify *statistical units*, which in turn would be employed in collecting and aggregating these data.

Having devised a map of *income expenditure preferences* of individuals and households in this fashion, a structuralist would then be in a position concretely to explore the empirical "fit" between data and models devised at both levels simultaneously. Moreover, since his or her method would necessarily involve an explicit mapping of phenomena at one level of organization into the other, the relationship between the two would not have to be left to guesswork—as it usually is in conventional social science.[49]

As this example suggests, then, one of the common themes in structural analytic research to date has been the exploration of phenomena going on at several different levels of detail within models. In order to study this, new concepts had to be devised because "classic" formulations of the problems of levels and units of analysis were too murky for the precise sorts of techniques, such as graph theory, which structuralists were using. The further refinement of these techniques, moreover, led to a transformation of the Durkheimian notion of *categories* of social and individual facts into a concept of systems with *layers* which are interdependent and, hence, interact with one another in complex ways. Thus, instead of a simple dichotomy between levels, structural analysts suggested that relevant phenomena might be occurring at more than two levels in a system, and that the boundaries between these layers, and the dynamics of the processes going on within and between them, might constitute proper areas for further research. As we shall see in later chapters, this has proved to be an extremely useful hypothesis.

Boundaries and Boundedness

The clarification of these "classic" problems of levels and units of analysis has been, to some degree at least, a by-product of structural research into a number of different but related issues. The most important of these has to do with the *boundaries* between groups and the internal and external conditions which maintain them.

In the same fashion that we may define boundaries between levels of detail in models of "real" systems, we can also specify boundaries between sets of closely related elements *within* each of these layers. Each set so defined corresponds to what is usually referred to as a social group. In cases where we choose simple social groups as units of analysis, the boundaries of "groups" and boundaries of "units" coincide.[50]

In most cases, however, more than one measure of "groupedness" is necessary for understanding a given phenomenon. This fact corresponds, intuitively, to the kind of *multiple* or *overlapping group memberships* that most of us experience if we live in complex urban centers. Thus, in some sense, a "professor" can be thought of as someone who belongs, simultaneously, to (a) a professional group (e.g., sociologists, economists, or architects), (b) a university bureaucracy ("the faculty of the University of Southern North Carolina at Boodle [USNC at B]"), and (c) a class ("middle class professionals" or, more self-servingly, "the Intelligentsia"). At the same time, of course, he or she could be a member of a range of other entities as well: ethnic groups, political groups or parties, religious groups, alumni bodies, or whatever.[51]

By combining these categories of memberships haphazardly, we could discover some intriguing sets of people, for example, "socialist Catholic professors of otology at USNC at B who are alumni of Harvard College." In both theory and practice, however, social scientists are only interested in social groups whose behavior or structural role they can interpret. Therefore, while in any situation there are a finite but large number of groups to which the members of a population may belong in principle, in practice social scientists deal with only a few types or combinations of these.

The reason for this is clear: since models are constructions of "reality," the definitions of relationships between objects which models imply must rest on some theoretical rationale. Consequently, the real task in model building is to find ways of theoretically interpreting the reduction or simplification of "reality" we can accomplish by imposing categories on populations of elements. Hence, to the extent that there is an upper limit on the number of sets into which we can divide a given population, it most directly reflects the richness of our conceptual apparatus, and not "reality" itself.

Thus far, most structuralists and nonstructuralists would agree. They would part company, however, over the issue of how one ought to go about specifying those operations to be used in simplifying "reality." Conventional—or, as they are sometimes called, "aggregative"—social scientists implicitly conceive of social categories as simple sets, that is, ones whose elements have no intrinsic relationships to one another. Hence, while the elements of a given set may be *ordered* in some manner, they must be treated as independent in other respects. This fundamental assumption underlies conventional conceptualizations of social science problems, whichever perspective—functionalist, neo-classicist, Marxist, structural functionalist, and so forth—a given analyst may choose to adopt or espouse.[52]

By contrast, structural analysts, as we noted earlier, derive definitions of groups from an analysis of patterns of relations among elements. This implies an entirely different logic. If, in simplest terms, the credo of conventional or aggregative social science is "categories have consequences," the structuralist rebuttal would be "consequences have categories," that is, patterns of relations among members of sets of elements produce or yield those social entities which we recognize and interpret as social groups. To the extent that we are able to create reasonable models of social "reality"—that is to say, to the extent that the relations in our models are good analogs for the effects of "real" elements on one another—networks of ties among them depict the interplay of general forces or processes, which then generate observably different behavioral patterns. The concrete intermeshing of these forces in the "real world" produces or creates groups of people or institutions which we recognize as such. Thus, when a social scientist lumps real-world objects together into social classes or strata, he or she is simply acknowledging the consequences of that set of social processes which structuralists have tried to model.

Since both conventional and structural methods eventually yield sets of like elements (which we label as groups), a superficial observer would see these two approaches as quite similar. Structurally specified groups, however, differ from conventional ones in an important way: by definition, the elements of those subsystems upon which structuralists base their definitions of groups are *not* independent. This apparently simple difference has important intellectual and technical consequences.

In any empirical case, the *patterns of ties* among the nodes of a network model of a social system may not perfectly coincide with, or *map over*, one another. In fact, when more than one type of tie is involved, they almost never do. Thus, some means has to be found of reconciling differences between measures of connectivity. In addition, even simple networks—ones that reflect only one type of tie—do not usually break down into "discrete" (i.e., nonconnected) sets of nodes. These, then, must be "cut" or subdivided in some consistent way. The boundary problem is, simply put, one of finding some way of doing this which preserves the internal coherence and relative external isolation of groups of nodes, or *clusters*, defined by a set of at least partially overlapping sets of ties among them.[53]

While graph theory provides us with a rigorous definition of one highly restricted type of cluster—namely, a *clique*—in which all nodes are directly connected (*adjacent*) to the others,[54] structuralists have been baffled by the problem of finding a mathematically rigorous formulation for the general case: there are almost as many clustering measures in the literature as there are people who use the idea. Consequently, cluster detection methods are usually most directly applicable to the particular empirical case a given researcher is studying at the time, and concrete "solutions" to the boundary problem are more plentiful than well-focused theoretical discussions of it. However, both aspects of clustering—the practical or technical problems of how one goes about doing it and the theories behind it—have constituted, as we shall see later, important and recurring themes in structural analysis.[55]

Transitivity of Ties or Indirect Effects

Most structural analysts have been deeply interested in some aspects of the dispersion of information in systems. These questions generally fall under the rubric of *indirect effects* or the *transitivity of ties*.

A simple binary relation between two nodes can, as we suggested earlier, be *composed* together with other relations in order to form extended or *compound* relations. Hence, structural analysts generally refer to the act of creating all the extensions of the patterns of simple ties among the nodes of a network as the *composition operation* of that network. In cases where it is possible to treat *higher orders* (t^N) of a particular type of tie as though they implied the same thing as a simple *first-order* tie, we can say that the type of tie in question is *transitive*.

This idea is most easily understood if we express it mathematically. Let us assume that there is a particular type of relationship—say, friendship—which is possible between members of a college dormitory. Let us refer to this relationship by the letter "R." If one person in the dorm, *a*, expresses friendship toward another person, *b*, we can say this symbolically by writing *aRb*. Literally translated, this "sentence" simply means "*a* expresses friendship for *b*." If *b*, in turn, expresses friendship for c, we can write this bRc. When the R relationship exactly corresponds to its extension, RR—that is to say, in this case, when "friends' friends are friends"—we can show this by writing R=RR. This implies that if aRb and bRc, then it is *always* true that aRc. When this condition is met, we say that the R relation is transitive.[56]

We can treat different types of ties in exactly the same way if we feel it is reasonable to compose them together. If we use the letter "A" to mean "forms an alliance with," then if aRb and aAb, we can write this: $a(RA)b$. If $b(RA)c$, as well, and the *compound* relation, RA, is transitive, then the statement $a(RA)c$ will always be true in a similar fashion.

This mathematical idea of transitivity corresponds to the observed tendency of elements in social systems to have *indirect* effects on one another. In the next chapter we will examine a number of examples of these in some detail. At this juncture, however, it is important to note that if the relations among the elements of a social system are highly transitive, this implies that an action by one of them on another is directly translated into effects that can be perceived or detected by all elements that are directly or indirectly linked to the receiving node in a similar way. The precise pattern of transitivity or lack of it (*intransitivity*) among the nodes in a network model of a social system, as a result, comes very close, by itself, to capturing precisely what social scientists have historically meant by the concept of "social structure."

Structural analysts have explored a range of social situations in which they can examine indirect effects of this kind with interesting and sometimes even startling results. The oldest continuing examples of this theme are studies of simple patterns of indirect contact or influence in populations. Since the 1930s, it has been known that the "volume" (i.e., number of contacts per person) of *direct* contacts between people (a knows b) varies enormously according to the social class, occupation, sex, age, and so forth, of the persons involved.[57] As long as someone's first order contact network (at t^1) is not largely self-enclosed—that is to say, as long as one's acquaintances do not merely know one another—this volume of acquaintances rises rapidly when one includes indirect effects, such as friends of friends. Indeed, in empirical studies where persons' first-order contacts vary from a few persons to a few thousand, their second- and third-order acquaintanceship soon reaches everyone within a large nation state![58] The more central a person is to a series of overlapping networks, the more rapidly this can occur. For example, a politician with a large number of first-order acquaintances and widespread contacts might, within a few jumps, reach most of the world's population.

We refer to empirical studies of acquaintanceship distances between people in large populations of this kind as "small world studies," after the omnipresent phenomenon of meeting someone new and discovering that you know someone in common ("Isn't it a small world!"). Stanley Milgram and his associates, following on earlier research by Ithiel de Sola Pool, Manfred Kochen, and others, conducted a series of empirical studies in the late 1960s and early 1970s in which they traced *paths* between a sample population and randomly selected *targets*.[59] Both the theoretical and technical issues involved in carrying out research of this kind have subsequently become major concerns in structural analysis.

The simple fact that people know one another, while of course interesting in its own right, tells us relatively little about the content or implications of their relationships. A rich vein of issues along these lines has been opened up by structuralists probing the consequences of contact networks for a number of ordinary *search processes* which people go through in order to find jobs, get help with family problems, locate an abortionist (then illegal), contact politicians, or find goods and services. Harrison White laid out theoretical dimensions of these "search" problems in two important articles—"Search Parameters for the Small World Problem" and "Everyday Life in Stochastic Networks"—and encouraged concrete studies (notably ones by Mark Granovetter, Nancy Howell, and Barry Wellman) which extended and developed these insights in specific substantive contexts.[60] Granovetter's research on the structure of job markets—especially as formalized and interpreted by another student of White's, Scott Boorman—Howell's study of women seeking abortions, and Wellman's work on the network basis of urban communities are widely regarded as landmarks in structural analysis and have contributed to a reorientation of conventional social inquiry along structuralist lines.[61] Highly transitive structures of this kind, combining, in many cases, a plethora of "weak" and "strong" ties, are now being discovered in a variety of contexts where they appear to be instrumental in integrating large-scale social structures and articulating group interests. New methods of examining these structures, and the new substantive insights into social process which they have provided, will be treated in detail later in this book.

Centrality

The large, weakly connected, and highly transitive networks which social scientists typically discover when they investigate systems of interpersonal ties are often difficult to interpret. Statistical measures— even ones which only deal with gross, overall properties of such networks—can be quite misleading. Analytic techniques which search for specific, small-scale structures within these networks are normally frustrated by the sheer *sparseness*, that is, low density, of connections between elements.[62] Thus, in order to make sense out of these intricate spiderweb-like patterns, structural analysts have had to devise or create theoretical constructs which reflect the specific properties of these new forms of data. The most important developed concept of this kind is the idea of the *centrality*, or relative central location, of nodes within a network.

In the late 1940s and early 1950s, researchers interested in group efficiency in problem solving and a number of related issues—for example, "perceptions of leadership," "personal satisfaction"—began to devise measures of the centrality of persons within small, task-oriented groups. Centrality, in this sense, was viewed as an *attribute* or individual characteristic of persons who played specific roles within given experimental groups.[63] As such, it was simply treated as a variable which could be correlated or combined, in a conventional way, with other, *aggregate* properties of nodes or elements.

The concept of centrality, however, was successfully borrowed and adapted by scholars with broader macrosocial interests (e.g., the coordination of diverse social processes, the efficiency of transportation or communications patterns, organizational structure). By the mid-1960s, both the conceptual foundations of the idea of centrality within networks and techniques for measuring it were sufficiently well understood to allow for its extensive use in analyzing a range of other important social phenomena: technological diffusion or innovation, intercorporate ties, systems of patronage or patron-client relationships, illicit markets, and community organization.[64]

The centrality measures most widely used in these studies, although limited in some ways, are also the most conceptually straightforward. Given a network in which each node is linked, directly or indirectly, to others at some order of t—what we call a *connected graph*—it is possible to specify a *path* (sequence of one or more relations) of some length between any two nodes. If there are paths between all these nodes in a network (which will always be true if the graph is connected), we can say that each is *reachable* from the others at some specifiable *path-distance*. Where any two nodes are directly connected to one another—that is to say, where they are directly reachable or are a distance of t^1 apart—we refer to them as *adjacent*.

Most centrality measures which have been successfully used in structural analysis are based on simple counts of the number of *other* nodes adjacent to a given node, that is, its *degree*. Intuitively, when an element (a person) is "close" to others, it is easily able to "get in touch" with them, find out what is going on, and pass on information. If an element is directly connected to a large number of others, we think of it as being at the "hub" of things: busily "taking calls," contacting other elements in the network, and shaping the process of information exchange. *Degree-based* measures of centrality of this kind thus make sense where we are trying to distinguish what *potential* the elements of a system have for communications activity.[65]

In all but a few cases, these simple counts, by themselves, will tell us most of what we want to know, namely, the sheer amount of communications activity in which elements may be engaged. The "degree" of any node, however, is obviously a function of the number of "others" to which it *can* be attached. This, in turn, is a function of network size (number of nodes). Thus, unless a researcher is dealing with small, bounded networks—such as task groups—it is necessary to standardize measures of the "degree" of nodes in some fashion.

The simplest, and most frequently followed, way to do this is to calculate a measure of the *relative degree* of a given point. Since, where there are N nodes in a network, any given one may be connected to $N - 1$ others, if we divide our simple count of the number of adjacent nodes by $N - 1$ we are able to construct a ratio of the actual or observed adjacency of that point to the potential number of first-order ties which could have been made. Structuralists often base notions of centrality on this measure of the *relative degree* of nodes where systems are large or where the opportunities for contact among elements are constrained by external cost factors.[66]

This operationalization of the concept of centrality is clearly inadequate, however, if we want to deal with the strategic implications of the position or location of elements within a network. Structural analysts, as a result, have constructed two additional types of centrality measures that are designed to reflect this.

The first of these measures focuses on the *shortest paths* (or "geodesics") linking nodes to one another. Given two nodes, a and b, any and all points falling on the only "shortest path," or on all equally short paths, linking these two nodes are said to fall *between* a and b. Whenever two nodes are not adjacent, but are separated by a point or a number of points, these "between" nodes can potentially shape (or distort) the flow of communication as it affects these two elements. By extension, whenever the "betweenness" of a node or set of nodes is high, this may be taken as a measure of the control which a node or set of nodes may exercise over the communications process as a whole and, under certain circumstances, over the behavior of other elements.[67]

The second type of strategic centrality measure rests on a generalization of the notion of adjacency. If any two nodes may be said to be adjacent if they are joined by a relation at length t^1, then we may conceive of points which are "second-order adjacent," that is, at length t^2, and so on. If we then sum the *orders* of interconnection between any given point and all other nodes within a network, in terms of shortest path distance, we have an index of the relative isolation or (as Linton Freeman suggests) "inverse centrality" of that point with respect to all others.[68]

During the last several years, structural analysts have clarified each of the notions of centrality represented by these measures. They have extended them, moreover, so as to provide indices of the overall centrality or *centralization* of graphs as a whole. As we shall see later on, the most methodologically sophisticated efforts of this kind have been based on work done by Linton Freeman and Phillip Bonacich.[69] In addition, extensive and, in many respects, ingenious use has been made of sophisticated centrality and centralization measures in describing and analyzing connections between corporations and state bureaucracies, in locating centers of power within financial networks, and in addressing a number of other important and otherwise intractable substantive problems.[70]

Small-scale Structures

Some networks, rather than being sparsely or weakly connected, are so dense that commonplace clustering techniques fail to discriminate adequately among sets of closely related elements. Hence, defining boundaries between clusters becomes partially, if not wholly, arbitrary. In addition, even where the overall density of a graph is low, it may be quite high in some local area. Most clustering techniques will detect these dense *neighborhoods*, but, given the generally sparse nature of the surrounding terrain, will not draw fine distinctions among elements falling within large and amorphous clusters.

Faced with these problems, structural analysts have created methods for identifying or *detecting* explicitly defined small-scale structures within a larger network.[71] These structures are then typically used to describe and interpret overall properties of the system which is being modeled. Since this cannot be done without having some clear means of interpreting these "mini-structures" in mind beforehand, structuralists have also been forced, under these circumstances, to lay out a detailed rationale which associates given structural patterns with larger theoretical dimensions of social process.

The clearest and best-developed examples of this approach have focused on three-way associations among elements called *triads*. Georg Simmel, one of the precursors of modern network-based studies of the "forms" of social structure or process, first noted the interesting strategic possibilities implicit in three-way, as opposed to two-way or *dyadic*, patterns of interaction.[72] Fritz Heider, a psychologist, was able to codify these notions into a theory of "cognitive balance" which explained the options open to actors in establishing reciprocal orientations in triadic groups.[73] Dorwin Cartwright and Frank Harary, in a seminal paper published in 1956, restated Heider's notion of "balance" in graph-theory terms and showed its implications for the organization of sets of actors into cliques.[74] James Davis extended this model further and showed that it was a special case of a more general structural model.[75] Davis and Samuel Leinhardt then forged a crucial link in the chain between small-scale structures and larger structural properties of systems by demonstrating how directed relations could eventuate in a network of hierarchically ordered cliques. [76]

With this preliminary work in place, the stage was set for both a general theoretical treatment of the relationship between small-scale events (sets of positive interpersonal ties) and the larger organization of a system, and the development of a method for surveying or enumerating the number of triadic relations within networks as a means of describing structure. In 1970, Paul Holland and Samuel Leinhardt, in a monograph entitled "A Unified Treatment of Some Structural Models for Sociometric Data," provided an exhaustive and theoretically acute statement of the relationship between small-scale and general organization within systems of ties. [77] This was then condensed and clarified in a paper published in the same year, "A Method for Detecting Structure in Sociometric Data," which developed a statistical interpretation of the occurrence of intransitive triads as a structural measure. [78] The real strength of Holland and Leinhardt's method lies in the fact that it allows for comparisons between the number of intransitive triads in actual or observed structural models, and a single statistic, T, which describes their distribution in random graphs. Given a measure of this kind, structural analysts were then able to explore "structure" in a range of substantive contexts, and to compare and generalize their results.

The Davis-Holland-Leinhardt method, of course, rests on a number of assumptions about the *morphology* of relationships and the *content* or meaning of ties. In particular, the strength of ties is assumed constant across any given network. This is a reasonable assumption, given the origin of their work in studies of interpersonal networks and affective relations. Structuralists, however, are also interested in situations in which the strength of ties may vary, and as a result cannot assume, a priori, that objects with the same abstract form will act in the same way vis-à-vis the larger system. Moreover, by extension, substructures based on similar principles, but with different shapes, may be functionally equivalent in this respect.

Stephen Berkowitz, Peter Carrington, Yehuda Kotowitz, and Leonard Waverman therefore adopted another approach to "detecting" small-scale structures within a larger network where they suspected that functionally equivalent clusters of nodes might be morphologically quite different. In 1975, they began studying the complex, weblike, and dense ownership and directorship connections among the members of a large set of corporations. [79] They were concerned with trying to detect "enterprises" (groups of firms operating under common control) in a context where the costs of forming new nodes ("firms") and relations (ownership ties and directorship interlocks) were low. Under these circumstances, given positive incentives toward the creation of separate legal entities ("firms"), we would expect them to proliferate. Berkowitz et al. suspected that this had happened and that these "enterprises" represented functionally equivalent "capital pools."

Their initial results indicated that the *patterns* of *simple* binary ties between firms were random and that if the strength of these connections was assumed constant, a clustering procedure would produce trivial results; that is, 80 percent of cases would fall into one cluster. They then devised a method of combining ties and gauging their relative strengths as indices of "control" among firms. A series of measures based on these procedures were then used in concert with a theoretically specified model of control to separate out the enterprises embedded in this densely connected web. Distributions of the sizes (number of firms) of enterprises were then calculated under each measure and compared using a standard inequality index (Gini).[80]

Both of these types of methods—ones which assume that tie-strength and morphology are constant and ones which do not—are relatively rare because they require some prior specification of the theoretical implications of particular patterns of ties. Hence, while the development of techniques for detecting small-scale structures is clearly an important theme in structural analysis, many researchers prefer to utilize methods, such as generalized clustering procedures, which do not require this kind of close specification of theoretical models at the outset. As we shall see later, both styles of work have important roles to play in the development of structural approaches to social inquiry.

GOALS OR OBJECTIVES

In this chapter we have briefly looked at the intellectual roots of structural analysis, its principal scientific orientations, and common themes in the work of its practitioners. Most of this discussion has been fairly abstract. In the chapters that follow, we amplify and extend each of the issues raised up to this point by directly tying it to a specific, concrete problem or set of problems in the analysis of some important aspect of social life. Each of the chapters that follow, as a result, begins by outlining a set of formal or analytic problems having to do with the patterned behavior of parts of a social system, and concludes by focusing on a series of substantive studies in which examining these issues has significantly improved social scientists' ability to model and interpret the "real" world.

Before proceeding with this discussion, however, it would probably be useful to clarify a few things. Despite its, in some ways, impressive beginnings, structural analysis is not a fully developed paradigm. Thus, each of the advances that will be described later raises, with it, some important unanswered questions which, in the final analysis, may be more significant than those that have been dealt with to date. This indicates, in some sense, that structural analysis is likely to be a fruitful way of looking at the world: a good paradigm should suggest new avenues for research. However, it is also frustrating: any sensible reading of the structural analytic literature leads to the conclusion that we have barely begun to probe its implications for the conduct of social inquiry. Thus, with rare exceptions, the results reported in what follows should be treated as tentative or indicative, and subject to revision due to further development of our knowledge in the area.

Structural analysis, as we have noted, approaches the task of describing or modeling and interpreting the social world in ways which are radically different from those adopted by conventional social scientists. Therefore, it is reasonable to ask whether, sometime in the future, structural analysis is likely to emerge as a separate field. In other words, do structural analysts intend to develop their area as a separate discipline intended to supplant the existing social science disciplines? The answer to this question is clearly no. While its emphasis on the structure of social systems and the systemic sources of behavior obviously falls outside the purview of conventional social science, the framework that structural analysis brings to bear on problems is clearly an extension of currents of research that have been flowing through the conventional disciplines for a long time. In fact, it is only in the last several years that social scientists doing structural research have come to recognize that their modes of addressing problems are categorically different from others. Therefore, it is far more likely, in the near term, that structural analysts will attempt to infuse their several fields with structuralist ideas and methods rather than hive off into a distinct academic ghetto.

In the final analysis, the principal goal of structural analysis as a field of research is *integrative:* to bring together and develop a set of perspectives and tools that will enable social scientists to delve more deeply into the systematic sources and consequences of social behavior in organized settings. This is not in any sense a new or revolutionary task: it is the central rationale of any social science that takes itself seriously. Structural analysts simply argue that the direction they have taken is likely to allow us to reach this objective.

Structural analysis, then, is best understood as an attempt to reintroduce "social science" to its fundamental roots as a specifically *scientific* enterprise. It explicitly rejects the argument that sees the social sciences as categorically different from others and, as such, pursuing a different logic, style of reasoning, different sorts of tools, and so forth. The chapters that follow are intended to show precisely how structural analysts have gone about trying to accomplish this.

ADDITIONAL SOURCES

Kemeny, John G., and J. Laurie Snell. *Mathematical Models in the Social Sciences*. Boston: Ginn, 1962.

Lazarsfeld, Paul F. (ed.). *Mathematical Thinking in the Social Sciences*. New York: Russell & Russell, 1954.

Leach, Edmund. *Claude Lévi-Strauss*. New York: Viking Press, 1970.

Olinick, Michael. *An Introduction to Mathematical Models in the Social and Life Sciences*. Reading, Mass.: Addison-Wesley, 1978.

2

Kin, Friends, and Community: The Structure of Interpersonal Communication

No area of study has been as closely wedded to individualistic assumptions, beliefs, and models as that conventionally known as "interpersonal relations." After almost a century of research into topics—such as families, peer groups, friendship, ethnicity, and urban neighborhoods—where this conceptual framework is only marginally useful, the predominant tendency is to continue to view the social world as a simple extension of the sentiments, motivations, and attitudes of individual actors. At best, conventional social science sees the ties which bind actors together into families, ethnic groups, friendship circles, and other social entities as products of individual consciousness or internal states of mind. At worst, social life is depicted as nothing more than the sentiments, motivations, and so forth, of individuals writ large. Sociologists and economists, moreover, have been as prone to think in this way as social psychologists or historians.

The reasons why this essentially eighteenth-century worldview has persisted long past its "time" are both profound and subtle. Individualistic models of the world are deeply embedded in the political and cultural apparatus of Western states. Our legal system, for instance, assumes "reasonable men" who rationally orient themselves toward a social world consisting of other "reasonable men." When an individual acts contrary to law, he or she is held accountable not only for his or her act itself, but for all the consequences which could reasonably be expected to flow from it.[1] As a result, even where the effects of an act are felt at three and four removes, Anglo-Saxon law holds the offending individual fully responsible. The bank robber who in making his getaway causes a traffic jam which, in turn, leads to a fatal accident is deemed guilty of first-degree murder, *not* manslaughter.

The same presumptions are implicit in most conventional social science models. Individuals, acting reasonably and rationally, are expected to be able to assume that other individuals will orient themselves toward them in the same fashion.[2] Where these expectations are not met, actors are seen either as pursuing some "limited" form of rationality or as behaving rationally in response to motives which are, themselves, hidden from the individuals in question. Thus, in the Freudian model, while the *basis* of "unconscious" motives may be irrational, the *means* chosen to display the individual's unconscious intent are readily understandable in rational terms.[3]

In models where the abstracted "rational actor" is *not* explicitly given, a kind of *methodological individualism* prevails. While in economics, for instance, the "rational optimizer" is only a construct reflecting the aggregate consequences of a series of individual choices, outcomes are the same as they would be if actors actually understood and were pursuing their own best interests.[4]

We can best appreciate how deeply these individualistic interpretations of reality have been etched into the worldview of conventional Western social scientists when we examine the vocabulary of motives which they attribute to other, non-Western peoples. For instance, Franz Boas's otherwise excellent explanation of the "potlatch" ceremony lays unusual stress on the *individual* rewards derived from it.[5] In the same vein, sociologists who have compiled detailed and, in many ways, sensitive "family histories" in Latin American *barrios* have often adopted the same individualistic analytic approach they would in surveying North American middle-class whites.[6]

These assumptions and models, moreover, have not been confined to the realm of pure theory. Social workers, psychiatrists, job counsellors, and school officials are trained to look for the sources of personal difficulties in individual feelings and attitudes and in their clients' consequent inability to "adjust" to others' expectations.

In both theory and practice, then, the strong tendency has been for conventional social science to associate personal dimensions of social life with individual characteristics. In this chapter we will examine how relaxing this assumption has enabled structural analysts to open this area to rigorous analysis in ways which have called into question some of the conventional literature.

Family and Kinship

Since the 1930s, the majority of sociological studies of "the family" have focused on the "nuclear family" (i.e., an adult male, an adult female, and their children) to the exclusion of other aspects of kinship systems.[7] This emphasis fits well with both a psychodynamic interpretation of the role or functions of families (e.g., nurturance, socialization), and with a form of methodological individualism in which actors create a stage upon which they act out a series of sociodramas based upon generalized conscious and unconscious motives and interests. It also neatly dovetails with a psychologistic view of the larger society as the simple product of individual encounters, reciprocal orientations, and interpersonal alliances.[8]

During the last two decades, however, structuralists have discovered a range of phenomena going on between and among nuclear family units which suggest that a larger, systemic focus is called for—even in studies of the dynamics of these smaller units themselves. In the mid-1950s, Elizabeth Bott and a group of colleagues connected with the Tavistock Institute in Britain conducted a series of in-depth "home interviews," "clinical interviews," and "case conferences" with 20 carefully selected nuclear families.[9] These interviews dealt with a wide variety of subjects, from highly specific aspects of the division of labor among, and social contacts of, family members to broad attitudes toward class, the social norms enforced within the particular social circles of the selected families, and deviance and conformity in general. In addition, "group discussions" were conducted under the aegis of a series of social, labor, and political organizations (townswomen's guilds, parent-teacher associations, community centre groups, and political-party branch meetings) in an attempt to situate the interviewees within the larger context of the community in which they lived.[10]

Initially, at least, the focus of these studies was on traditional issues in the organization of nuclear family life: task differentiation among, and role segregation of, men and women, modes of family decision making, emotional attachments and dynamics among family members, and so on. Bott in particular, however, was also interested in the degree of conformity between these internal features of nuclear-family life and those sorts of "social facts" which had traditionally been the object of sociological analysis since Durkheim's formulation of the levels problem: occupation, social class, the social division of labor, sex roles, and so forth. This led to what is best described as a tug-of-war between sociologically oriented and psychologically oriented members of the research team. The former preferred to look for the sources of differentiation of *conjugal roles* in the external categories or classes of actors to which family members might belong. The latter, including psychoanalysts and psychologists, preferred emphasizing individual variables and traits.

The outcome of this conflict was an analytic design—apparently worked out by Bott and A. T. M. Wilson—in which the rigid separation of events into "social" and "nonsocial" levels was discarded and an attempt was made to explore both the impact of *extrafamilial* processes and organization on *intrafamilial* patterns of organization and the more conventional kinds of intrafamilial dynamics which had usually been examined in clinical case studies of nuclear families. As time went on, the distinctions between these two "realms" began to blur even more and an attempt was made to integrate sociological and psychological theory.[11]

As a result, the work of Bott and her colleagues constitutes one of the first examples of the transformation of the classical distinction of levels of analysis into one between *layers* or boundaries within systems. While this distinction was not formalized as such, in this sense their work clearly represents an important step toward a more systemic and relationally oriented approach to social inquiry. Their results, moreover, unequivocally show the analytic power which studies of this kind can introduce into even very traditional research areas.

First, Bott and her colleagues established that there was a moderately high degree of interconnection between nuclear families in the modern, industrialized settings they examined. This ran counter to most sociological and much anthropological research during the period, in which it was *assumed* that a high degree of separation of family units into households had automatically induced a commensurate breakdown in kin organization. The realization that this was not true led other researchers to begin studying the concrete *patterns* and *dimensions* of the ties among kinsmen within these settings and, as we shall see later, to explore their larger functions within nontraditional societies.[12]

Second, they discovered that there was considerable variation in the extent to which people with t^1 ties to their sampled families were connected to one another independently. Bott et al. used this as a measure of the relative degree of *cliquishness* or closure of the networks of family members.[13] Where cliquishness was high (*close-knit* groups), the degree of segregation of conjugal roles (i.e., the extent to which the activities of husbands and wives are different and carried out separately) was also high. Where cliquishness was low (*loose-knit* groups), role segregation appeared to be low as well. This suggested that in more closely knit networks, there is a greater tendency to reach a consensus on norms and to be able more effectively to exert informal pressures toward conformity on members. Hence,

> If both husband and wife come to marriage with such close-knit networks, and if conditions are such that the previous pattern of relationships is continued, the marriage will be superimposed on these pre-existing relationships, and both spouses will be drawn into activities with people outside their own elementary family (family of procreation). Each will get some emotional satisfaction from these external relationships and will be likely to demand correspondingly less of the spouse. Rigid segregation of conjugal roles will be possible because each spouse can get help from people outside.[14]

Finally, all things being equal, networks of kin-ties tend to play a more important role in the after-marriage lives of women in highly role-segregated families than they do in the lives of their husbands. Colleagues and friends, by contrast, tend to play a commensurately greater role in the networks of these men. Since kinship systems tend to be more closely knit than networks of friends, there is a tendency for men in these nuclear families to be drawn into their wives' kinship systems and for these kinship systems to play a commensurately greater role in the orientation of these families than the males' social contacts. In intermediate cases or in "joint" marriages, neighbors tended to form a larger proportion of the social contacts of both husbands and wives, and wives' kin had comparatively less impact on the nuclear-family relationship.[15]

While the conclusions reached by Bott and her co-workers were highly tentative at the time, other research has tended to bear them out. Michael Young and Peter Willmott, for instance, obtained consistent results from a comparative study (conducted at approximately the same time) of kinship in an inner-city borough (Bethnal Green) in East London and a post-World War II "housing estate" built by the London County Council on the outskirts of the city.[16] In Bethnal Green, couples tended to begin married life in the wife's parents' home. This reinforced the tendency toward closer associations between the couple and the wife's kinsmen which Bott and her colleagues observed. Even in those (rare) cases where couples began marriage in the *husband's* parents' domicile, however, wives continued to maintain close contacts with their mothers or other female kin.[17] Later in the marriage, couples tended to set up housekeeping in separate dwellings. Once again, however, the strong propensity was to maintain close ties to the wife's kin and to find housing close to them.[18]

While kin-ties in Bethnal Green appeared to be closest between mothers and daughters, the pattern of maintaining closer social contacts with the wife's family generalized to other kinsmen, both male and female. Thus, Young and Willmott report that in their sample 42 percent of the brothers of wives had been in contact with the sampled nuclear families during the previous week, as opposed to 27 percent of husbands' brothers. Similarly, 52 percent of wives' sisters had been contacted, as compared to 35 percent of husbands' sisters.[19]

While Young and Willmott made far less systematic use of network-based concepts and methods in their research than Bott and her colleagues did, the fact that Young and Willmott examined two radically different kinds of neighborhoods allowed them to observe the independent impact of broadly *ecological* factors on those aspects of kinship which they were able to deal with in detail. In "Greenleigh," the suburban housing estate they studied, former Bethnal Greeners experienced a marked decrease in social contacts with kinsmen. This decrease in the average number of weekly contacts with relatives, moreover, was more pronounced for women than for men. Before migrating to "Greenleigh," men averaged 15.0 contacts per week with their own or their wife's close kin. After the move, they averaged 3.8. Women, who before moving had averaged 17.2 contacts, now averaged 3.0. Two years later, the averages for men and women in the follow-up sample were 3.3 and 2.4, respectively.[20]

Young and Willmott stress the importance of sheer geographic distance in altering the intensity of ties between members of the nuclear families living in "Greenleigh" and kin left behind in Bethnal Green. Thus, husbands who continued, on balance, to work in Bethnal Green, maintained slightly more contact with kinsmen than wives who, for the most part, did not.[21]

From a structural point of view, this and related research undertaken by Young and Willmott[22]—though methodologically restricted by its failure to examine second-order and third-order contacts—strongly suggests important analytic dimensions which were not present, to any significant extent, in Bott's work. First, while Bott noted that families which moved tended to lose social contacts with kin, Young and Willmott explicitly describe a *process* in which geographic mobility leads to a loosening of kin-ties, greater reliance of husbands and wives on each other, and a consequent breakdown in role segregation within the home. This is important because structuralists are ultimately more interested in uncovering the dynamic properties of kinship systems, and their relationship to the environment in which they occur, than they are in simple descriptions of patterns of connectivity.

Second, Young and Willmott note that while labor exchanges and unions were beginning to take over much of the task of locating jobs and placing people in occupations that had traditionally been allotted through informal networks in Bethnal Green, these ties still played a considerable role in integrating nuclear families into the work world and, indirectly, into the economic system generally. Due to their interest in the *normative* integration of nuclear families into the community, Bott and her co-workers had not focused on these kinds of potential uses of interpersonal networks. For structuralists, however, the *multiple* functions which ties of this kind may perform is an important clue to the relationships among the processes which they may be modeling: a concrete association between two or more functions suggests that the processes underlying these systems of connections may be the same or similar. Young and Willmott were thus suggesting a fruitful set of hypotheses which could later be tackled directly by other scholars.

Finally, in one particularly interesting chapter in *Family and Kinship in East London*, Young and Willmott describe the way in which the migration of families from Bethnal Green to "Greenleigh" was facilitated and orchestrated by networks of kin-connections. This notion of a network-based process underlying patterns of migration—what we now refer to as a *migration-chain-effect*—has subsequently become one of the most important areas in structural analytic research.[23]

Friendship and Other Nonfamily Ties

In the early to mid-1960s, another group of British-based researchers began to publish studies on the organization and impact of interpersonal ties which took up many of these same themes. As was the case with Young and Willmott and Bott, they had been inspired by John Barnes's path-breaking use of networks as descriptive tools in field research.[24] Unlike Bott and Young and Willmott, however, their work demonstrated a sustained interest in the morphological properties of personal networks and, thus, more closely reflected Barnes's *analytic* intent, namely, demonstrating the relationship between network morphology and social behavior.

Adrian Mayer, for instance, traced the chains of influence used by candidates in mobilizing support during an election campaign.[25] The shape of these networks—and hence their ability to transmit or communicate information—appeared to play a decisive role in determining how successful the process of influencing voters was. Bruce Kapferer analyzed a similar process of mobilizing support through *egocentric networks*, that is, ones in which all nodes are first defined at t^1 to an initial node or "starter," and then all interconnections *among* this set of nodes are completed.[26] In his case, however (a study of the "Cell Room," or processing plant, of a mine in Zambia), *strength* of ties, as opposed to simple transitivity, played an important role in determining both the morphological properties of systems of interpersonal relations and the outcome of the competition for support by two disputants. Prudence Wheeldon examined another case in which an established group used pressure through both kinship and friendship ties to thwart a challenge to their leadership in a formal voluntary association. Here, the strategic location of the established group and the high degree of overlapping in the patterns of ties of "challengers" and the "establishment," taken together, determined how effectively this could be done.[27] Peter Harries-Jones reported on research in which *putative ties*—that is, ones which are commonly believed to exist—as well as clearly demonstrable kin connections, were used to form and maintain a political organization.[28] And David Boswell examined cases in which kinship and tribal co-membership were *differentially* "accessed" by people to mobilize help in personal crises.[29]

Each of these studies, while more descriptive than analytic in intent, explicitly used graph theory to characterize and interpret the form or pattern of observed relationships. Most earlier research had not done this. As a result, it did not always clearly distinguish between (a) properties of network models of "reality" which were due to the *mapping rules* used to translate observed relationships into graphs and (b) those which arose from systematic differences in actual patterns of interconnection. In earlier fieldwork studies, consequently, different results would crop up which were simply accidental by-products of the use of slightly different graphic techniques. However, J. Clyde Mitchell collected together many of the representative papers written during this second wave of field studies and was able to provide an overarching framework in which methodological and substantive disagreements could clearly be distinguished from one another. In the process, he also built bridges to the more formal interpretations of networks which were being developed in the United States at the time.[30]

Since the settings in which these second-wave studies were conducted were relatively small and bounded (e.g., a work group, a voluntary organization), analysis could still largely be carried out by *inspection*, or, more colloquially, "eyeballing" of the data. This is not as easy to do, however, either where patterns of interconnection between nodes are more complex or where a large number of nodes are involved. In addition, while second-wave researchers intuitively understood that different approaches to collecting and interpreting network data would, in themselves, sometimes yield different results, they had no theoretical model which could explain why this would happen.

STRUCTURAL MODELS OF SYSTEMS OF INTERPERSONAL TIES

Before taking up more recent structuralist studies of interpersonal relations, let us clarify some of these methodological issues. Up to this point, when we talked about nodes, relations, graphs, patterns of ties, and so on, it was not necessary, in most cases, to do this very formally. Most people intuitively grasp these concepts without actually having to visualize a network in detail. This approach is more difficult to sustain, however, when dealing with subtle differences in the ways in which these ideas may be used.

Basic Concepts

As we noted earlier, networks consist of nodes and relations. We conventionally represent a "node" by a single point in a graph, or by a circle or box corresponding to it. "Relations" are usually shown as lines joining these nodes—with different kinds of lines depicting different types of ties. Thus, a small system in which two friends (nodes) "like" one another (relations) could be graphically represented as in Figure 2.1.

Figure 2.1

Friendship Ties ("Liking") between Two Nodes, α and β

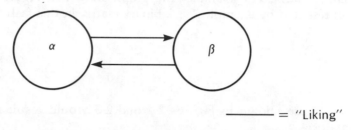

———— = "Liking"

If these same persons were also kinsmen, we could show this by simply adding a relation to the existing graph as in Figure 2.2. Note that in Figure 2.2 relations with arrowheads imply direction; for example, α likes β and β likes α. In this case, *kinship* is defined "nondirectionally." But an *undirected* tie of this kind cannot be practically distinguished, on this basis alone, from a situation in which both a given tie and its *reciprocal* (that is, the same tie, but opposite in direction) occur simultaneously. Thus, the graph shown in Figure 2.2 can be simplified, as in Figure 2.3.

Figure 2.2

Friendship ("Liking") and Kinship Ties between Two Nodes, α and β

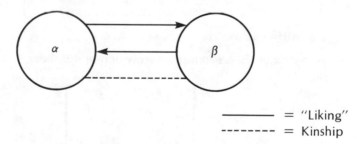

———— = "Liking"
-------- = Kinship

Figure 2.3

Simplified Version of Figure 2.2

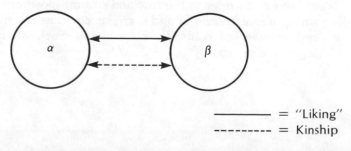

———— = "Liking"
--------- = Kinship

In most cases, it is therefore simply a matter of convention whether or not we use arrowheads in depicting *implicitly* reciprocal relations. Assuming all things constant, both forms of the graph will yield the same matrix. For example, the graph shown in Figure 2.1 could be represented by a simple 2×2 binary matrix in this fashion:

$$
\begin{array}{c|c|c|}
 & \alpha & \beta \\
\hline
\alpha & 0 & 1 \\
\hline
\beta & 1 & 0 \\
\hline
\end{array}
$$

Those shown in Figures 2.2 and 2.3 would result in two binary matrices:

Friendship Matrix   Kinship Matrix

$$
\begin{array}{c|c|c|}
 & \alpha & \beta \\
\hline
\alpha & 0 & 1 \\
\hline
\beta & 1 & 0 \\
\hline
\end{array}
\qquad
\begin{array}{c|c|c|}
 & \alpha & \beta \\
\hline
\alpha & 0 & 1 \\
\hline
\beta & 1 & 0 \\
\hline
\end{array}
$$

In each of these cases, "rows" correspond to "senders" of ties and "columns" to "receivers." If we wanted to combine the friendship and kinship matrices shown above into one matrix indicating what Mitchell refers to as *multiplexity*, that is, the existence of multiple ties between nodes, we could simply sum these binary matrices together using the rules for matrix addition:

Definition 1: Matrix

A matrix, A, is a rectangular array of real numbers $A = [a_{ij}]_{m,n}$, such that

$$
A = \begin{bmatrix}
a_{11} & a_{12} & . & . & . & a_{1n} \\
a_{21} & a_{22} & . & . & . & a_{2n} \\
. & . & . & . & . & . \\
. & . & . & . & . & . \\
. & . & . & . & . & . \\
a_{m1} & a_{m2} & . & . & . & a_{mn}
\end{bmatrix}
$$

where i and j refer to the row and column positions of an element within a matrix, and m and n are specified as the number of rows and number of columns, respectively. Such an array is called an $m \times n$ matrix.

Definition 2: Matrix Addition

Let $A = [a_{ij}]_{m,n}$ and $B = [b_{ij}]_{m,n}$ be two matrices of the same size. Then $A + B$ is the matrix $C = [c_{ij}]_{m,n}$ where

$$c_{ij} = a_{ij} + b_{ij}$$

for every i, j pair, $i = 1, 2, \ldots , m$ and $j = 1, 2, \ldots , n$.

To illustrate, if we designate the friendship and kinship matrices derived from Figures 2.2 and 2.3 as matrix A and B, respectively,

$$c_{11} = a_{11} + b_{11} = 0 + 0 = 0$$
$$c_{12} = a_{12} + b_{12} = 1 + 1 = 2$$
$$c_{21} = a_{21} + b_{21} = 1 + 1 = 2$$
$$c_{22} = a_{22} + b_{22} = 0 + 0 = 0$$

In matrix form, C would be shown as

$$C = \begin{bmatrix} c_{11} & c_{12} \\ c_{21} & c_{22} \end{bmatrix} = \begin{bmatrix} 0 & 2 \\ 2 & 0 \end{bmatrix}$$

The matrix, C, resulting from the matrix summation of A and B, could then be depicted graphically by a *single multiplex network*, as shown in Figure 2.4. Note that combining ties through simple matrix addition abolishes distinctions between them. It thus implies a theoretical statement: if the matrices corresponding to two types of ties, A and B, are summed together (using simple matrix addition) into one which corresponds to a multiplex graph, then each $a_{ij} \in A$ and $b_{ij} \in B$ is substitutable for the other.

Figure 2.4

Multiplex Graph Showing Effect of Combining Kinship and Friendship Matrices by Matrix Addition

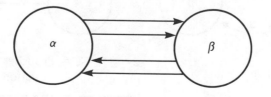

———————— = Kinship/Friendship Tie

This example illustrates why it is important to be very careful in choosing the rules used in translating observed relations in the "real world" into graphic form: slight variations in these will produce radically different results. Moreover, if we interpret two types of ties differently, this difference must obtain in the matrix-form representation of these ties as well. Many first-wave and second-wave studies did not do this consistently.

For example, we *can* draw a distinction between the *emanations* of nodes—that is, properties or characteristics of nodes which are projected onto others—and genuine *bonded-ties* between them. If α likes β, the relation between them is, thus, an emanation of α. α has reported to us about his or her projection with respect to β. This type of relation is qualitatively different from one which does not involve a *transmitted* property of α. If α and β were married, for example, the connection between them would suggest an overarching tie: α does not "send marriage" to β, and vice versa. Similarly, if we define a tie to mean that those joined by it have formed a partnership, this partnership is a bonded-tie which binds in or constrains (to use Durkheim's terms) the actors. It is not something which originates in one or both of them.

In practice, this distinction can be very subtle, indeed. For instance, let us assume that we want to represent the process by which a group of merchants amalgamates together into a larger consortium. First, one merchant proposes a merger to another. This clearly implies an emanation: the first merchant projects his interest in a merger onto the other. The other then reciprocates. Later the two are joined by some overarching bonded-tie, namely, partnership. The translation rules used in this case must be able to depict the difference between the two situations. Figure 2.5 shows one way this might be accomplished.

Figure 2.5

Initiatory and Final Phases in the Formation of a Partnership
between α and β

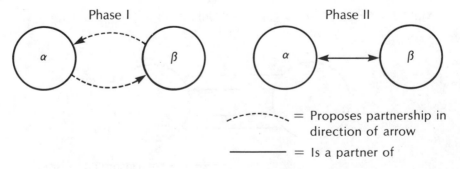

In Figure 2.5 a curved and dashed line indicates an emanation, and a solid line a bonded-tie. As we noted earlier, however, both of these types of ties would normally be represented by identical entries in their respective matrices. To wit:

Phase I Matrix   Phase II Matrix

$$
\alpha \begin{array}{c} \\ \alpha \\ \beta \end{array}
\begin{array}{|c|c|} \hline 0 & 1 \\ \hline 1 & 0 \\ \hline \end{array}
\qquad\qquad
\begin{array}{c} \\ \alpha \\ \beta \end{array}
\begin{array}{|c|c|} \hline 0 & 1 \\ \hline 1 & 0 \\ \hline \end{array}
$$

Thus, in order to analyze the process of merger, we must find some mathematically recognizable way of distinguishing between the reciprocal, but independent, emanations represented in the Phase I Matrix and the implicitly reciprocal relations shown in the Phase II Matrix. Structural analysts have devised two general ways of doing this.

The first of these takes advantage of the fact that bonded-ties are always reciprocated. Each directed tie, in effect, implies its mirror image: if α and β are nodes in a linear graph, and R is a binary relation between them:

$$\alpha R \beta \quad \text{iff} \quad \beta R \alpha$$

Since, by contrast, emanations are not inherently symmetric, it is possible to dissociate them from their reciprocals by introducing finer distinctions into their definition and then creating separate matrices based on these discriminations. In the present case, if we distinguished between the "proposes partnership" ties in the Phase I Matrix on the basis of the *time* when a proposal was made, in all likelihood we could generate two new types of ties, namely, initiating ties and responding ties. Similar discriminations might be drawn between them using the strength, duration, and so forth, of the ties as a sorting criterion. By systematically drawing finer and finer distinctions, up to the limits of parsimony, it would be possible to break down the "apparent" or "false" reciprocity of ties in the original matrix. In this case, of course, since two ties are involved, only one discrimination would be necessary.

The second strategy for distinguishing between emanations and bonded-ties requires more careful model building, but it is scientifically "cleaner." Hypothesizing that one of these types of ties is theoretically more important than the other is tantamount to arguing that it takes precedence over it. If we argued, for instance, that more permanent bonded-ties took precedence over more transient emanations, we could represent this by constructing a mathematical statement of the relationship between the two which was based on this assumption.

While the particular way this "precedence" can be built into an analysis would depend on the substance of the phenomenon we were trying to interpret—and we do not have enough information here to construct a precise model for our merchant-network example—the general approach structural analysts have followed is clear. In *matrix multiplication* the order in which terms appear is important.

Definition 3: Matrix Multiplication

Given $A = [a_{ij}]_{m,n}$ and $B = [b_{ij}]_{n,p}$, then the product, AB, is the matrix $C = [c_{ij}]_{m,p}$ where the entry c_{ij} of C is the result of the multiplication of the ith row vector of A by the jth column vector of B and $c_{ij} = a_{i1}b_{1j} + a_{i2}b_{2j} + \ldots + a_{in}b_{nj}$. Thus, $AB \neq BA$. For example, given two binary matrices,

$$A = \begin{bmatrix} 0 & 1 \\ 1 & 0 \end{bmatrix} \qquad\qquad B = \begin{bmatrix} 1 & 1 \\ 0 & 0 \end{bmatrix}$$

then

$$AB = C = \begin{bmatrix} 0 & 0 \\ 1 & 1 \end{bmatrix}$$

and

$$BA = C' = \begin{bmatrix} 1 & 1 \\ 0 & 0 \end{bmatrix}$$

By placing a given set of matrices, representing particular sets of relations, into a model in which they will be multiplied together, then, it is possible to establish an *order of precedence* among the *dimensions* which they represent. The extent to which any given matrix of ties will dominate the others will then depend on precisely how the model specifies that this multiplication of terms is to occur.

Properties of Systems of Interpersonal Ties

In the course of experimenting with different tools for representing "real" and hypothetical data, structuralists became aware of a number of systemic properties of graphs and matrices which had clear implications for the study of systems of interpersonal ties.

The first of these involved simple extensions of matrix multiplication. As we noted earlier, given a matrix, A, it is possible to calculate powers of this matrix which correspond to extensions or extrapolations of the pattern of the ties given in A, for example, AA or A^2, AAA or A^3. Given $A = [a_{ij}]_{m,n}$, then,

$$\text{where } A = \begin{bmatrix} 0 & 1 & 1 & 0 & 0 \\ 0 & 0 & 0 & 0 & 0 \\ 0 & 0 & 0 & 1 & 1 \\ 0 & 0 & 0 & 0 & 0 \\ 0 & 0 & 0 & 0 & 0 \end{bmatrix}$$

it would correspond to the directed graph (or digraph) shown in Figure 2.6.

Figure 2.6

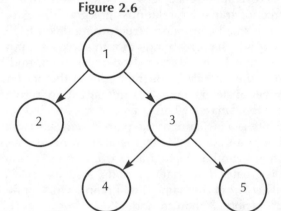

In practice, this digraph might be used to represent the command structure of a small organization, with the directed ties indicating the direction of flow of instructions. Squaring this matrix, following the rules of matrix multiplication, would yield:

$$A^2 = AA = \begin{bmatrix} 0 & 1 & 1 & 0 & 0 \\ 0 & 0 & 0 & 0 & 0 \\ 0 & 0 & 0 & 1 & 1 \\ 0 & 0 & 0 & 0 & 0 \\ 0 & 0 & 0 & 0 & 0 \end{bmatrix} \begin{bmatrix} 0 & 1 & 1 & 0 & 0 \\ 0 & 0 & 0 & 0 & 0 \\ 0 & 0 & 0 & 1 & 1 \\ 0 & 0 & 0 & 0 & 0 \\ 0 & 0 & 0 & 0 & 0 \end{bmatrix} = \begin{bmatrix} 0 & 0 & 0 & 1 & 1 \\ 0 & 0 & 0 & 0 & 0 \\ 0 & 0 & 0 & 0 & 0 \\ 0 & 0 & 0 & 0 & 0 \\ 0 & 0 & 0 & 0 & 0 \end{bmatrix}$$

The matrix resulting from this operation would correspond to the digraph shown in Figure 2.7. Note in Figure 2.7 that nodes 2 and 3 in Figure 2.6 "drop out" when we calculate the second-order indirect (t^2) linkages. The third power of the A matrix (A^3) would be zero-filled because no ties $\geq t^3$ are present. (All higher powers would, of course, be zero-filled as well).

Figure 2.7

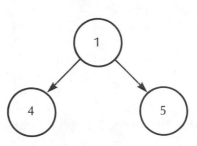

This correspondence between the powers of a matrix and the length of indirect or transitive relations in a graph is extremely useful. In the case of the small command structure modeled in Figure 2.6, by raising the matrix which corresponds to it to successive powers it is possible to discover what Frank Harary, Robert Norman, and Dorwin Cartwright refer to as the *reachability* of the nodes at the bottom of the hierarchy:[31] the number of steps it takes (t^N) an instruction from node 1 to "reach" his or her subordinates. Thus, as we mentioned in the first chapter, reachability is a measure of the path-distance between (adjacency of) the elements in a system. If, for instance, a node—such as node 3 in our illustration—can be reached at t^1 from node 1, its reachability vis-à-vis node 1 is greater than that of a node, such as node 4, which can only be reached at t^2 from this same initial point. Thus, node 1 is, in some sense, "closer" to node 3 than to node 4.

If the reachability of all elements is calculated, this information can be used to interpret a number of interesting characteristics of a system of interpersonal ties. First, if the range of reachability scores across a system is large, it suggests that there is a significant difference in the facility with which individuals can send or receive information. If a given network in which this were true were being used to model personal or collegial ties among a group of scientists, for instance, we would expect their field or specialty to develop very unevenly as a result of this pattern. Since publication of results usually follows research by several years, a scientist who was not well "plugged in" or well connected to the informal research network could easily be working on problems which had already been solved. Moreover, if he or she had completed a novel piece of research on his or her own, the results of this study would not be available for others to use in their work.

Second, if the distribution of reachability scores of nodes is skewed, it means that the ability of members to mobilize the resources implicit in the system is likely to be skewed as well. The distribution of reachability, in this sense, is implicitly an operational definition of *power;* understood as the ability to mobilize a system as a whole toward specified ends.[32] In systems where reachability is unequally distributed among members, we would expect, as a result, a commensurate inequality in the distribution of valued goods.

Finally, when the reachability of most nodes within a system of interpersonal ties—such as a "friendship" net—is relatively low, and assuming some rate of "error" linearly associated with transmission distances, we would expect misinformation or "rumor" to proliferate. Under these circumstances, the integrity of the system—its ability to retain or transmit information loyally—would be likely to be low as well, and to decline over time.

Harary, Norman, and Cartwright proposed another measure, *compactness,* which captures some of the flavor of all of these aspects of reachability simultaneously. The compactness of a network is, simply put, the proportion (or average proportion) of its nodes which can be reached from any point within it at a given distance. It thus reflects, as Mitchell notes, two facts: (a) the proportion of nodes which are reachable at all, and (b) a summary measure taken on the reachability scores of its nodes. Both of these aspects of the measure, of course, can be calculated directly by raising a given matrix to successive powers, in the fashion that we noted earlier.[33]

The second property of graphs or matrices that held out interesting implications for the interpretation of networks of interpersonal ties was the phenomenon known as *duality.* If we are given a graph consisting of two sets of objects—a set of nodes or points and a set of relations or edges mapped onto them—the *dual* of this graph is simply another graph in which the role of these objects is reversed: the nodes in the first graph become relations in the second, and the relations in the first become nodes in the second. Figures 2.8a and 2.8b present an example of this transformation.

Figure 2.8a   **Figure 2.8b**

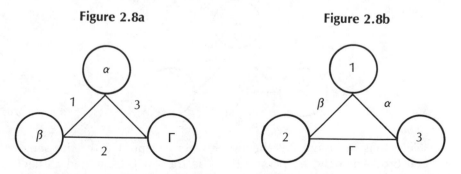

Figure 2.8a depicts a network in which three nodes, α, β, and Γ, are joined together by three distinct types of relations (or three distinct

instances of a given type). For example, if α, β, and Γ were households, relations 1, 2, and 3 might be defined in terms of people who had, at one time or another, been members of these households. Figure 2.8b shows the dual of this network: nodes 1, 2, and 3 would be people, and α, β, and Γ would correspond to "household memberships." Thus, for instance, in Figure 2.8a, the tie between α and β indicates that these two households are "joined" by the fact that person 1 has been a member of both. In Figure 2.8b, persons 1 and 2 are connected by the fact that they have both been members of household β. These duals represent, in effect, different dimensions of social structure. In this case, Figure 2.8a presents a model of *household structure:* aspects of this graph—the reachability of nodes, compactness, and so forth—would thus tell us about how integrated a set of households were with one another. Its dual in Figure 2.8b reflects the structure of *individual* relations or, hypothetically, intimacy: its linkage properties would be based on the extent of shared experience, distance between people, and so forth.

By comparing these two graphic forms of the same data, and their corresponding matrices, it is often possible—as Ronald Breiger noted in his paper, "The Duality of Persons and Groups"—to uncover otherwise unnoticed aspects of the interdependence of the organization of social groups and the processes through which individual ties are formed.[34] If one of the "households" in our example, for instance, had been a college fraternity, its unusual centrality might have provided us with a clue as to how other attachments in the graph had been formed.

Another aspect of duality also has useful implications for structural studies of interpersonal relations. If two graphs are isomorphic to one another, their duals need not be. Figure 2.9 shows a graph which is isomorphic to that presented in Figure 2.8a.

Figure 2.9

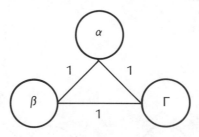

Note that in Figure 2.9 the tie between α and Γ is of the same general type—namely, a "person-tie"—as those we examined earlier. However, it is made by a different individual. Figure 2.10 depicts its dual.

Figure 2.10

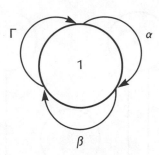

While the networks shown in Figures 2.8a and 2.9 are isomorphic, those in Figures 2.8b and 2.10 clearly are not. As these diagrams illustrate, there is no straightforward relationship between a given network, its isomorphs, and their possible duals.

This second aspect of the duality problem suggests a number of substantively interesting things. In Figure 2.10 the "self-looped" relations indicate that node 1 "shares membership" in the α, β and Γ households with him/herself. If we assume, along with Georg Simmel, that an individual's social identity is shaped by the number and types of social contexts in which he or she may be involved during the course of his or her lifetime, then given a network consisting of "contexts" (e.g., groups, households, cities) linked by person-ties, the self-loops in *its dual* may be used to construct a measure of the contribution of these contexts to each person's "identity." Figures 2.11 and 2.12 depict a slightly more complicated example of this.

Here, three sets of institutions—colleges, graduate schools, and engineering firms—are tied together by specified individuals who were associated with them during some part of their careers. Figure 2.12 shows the dual of this network.

Note that in Figure 2.12 each person (node) "shares" three of eight possible *self-looped* relations with him/herself. Each of these relations, in turn, corresponds to an institution with which a given individual has been affiliated during each of three stages in his or her *career trajectory*. The ties between individuals, by contrast, model the points of convergence of careers and, as such, provide a means of assessing the degree to which individuals might be expected to share aspects of their social identity with others.

Structuralists usually refer to these self-looped ties as *identity relations*. In most social network models they are assumed to be implicit in the graph (i.e., each node is understood to share all its ties with itself) and, hence, are not actually drawn. We can calculate the number of such relations in a straightforward fashion, however, because in an $n \times n$

Figure 2.11

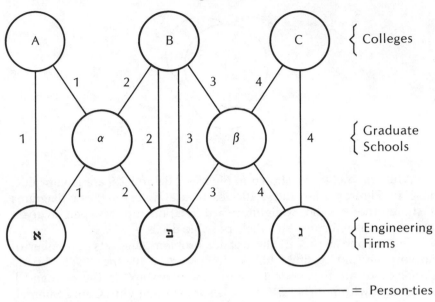

—— = Person-ties

Figure 2.12

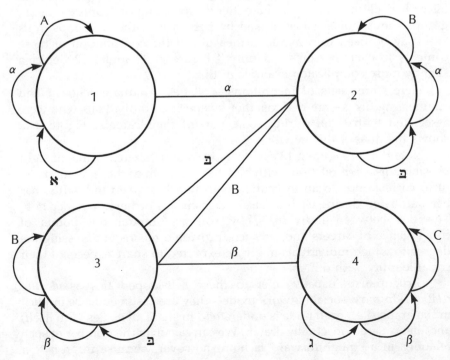

matrix, the entries corresponding to identity relations always appear on what is called the *main diagonal*.

Definition 4: Identity Relation

As used in social network analysis, an *identity relation* is an implicit or explicit tie which a given node "shares" with itself.

Definition 5: Main Diagonal

For a square matrix $A = [a_{ij}]_{n,n}$, the elements $a_{11}, a_{22}, \ldots, a_{nn}$ form its main diagonal.

Hence, where

$$A = \begin{bmatrix} a_{11} & a_{12} & . & . & . & a_{1n} \\ a_{21} & a_{22} & . & . & . & a_{2n} \\ . & & . & & . & \\ . & & & . & & \\ . & & & & . & \\ a_{n1} & a_{n2} & & & & a_{nn} \end{bmatrix}$$

all entries falling along the diagonal from a_{11} to a_{nn} would constitute the main diagonal of this matrix.

If we accept the idea that there is a direct relationship between the number of institutional environments in which a given individual has participated and the breadth of his or her social experience, then we can construct a systemic index of *context diversity* for any given person (node) in this *dual graph* by simply counting the number of entries ≥ 1 falling on the main diagonal of the corresponding matrix. The mean, mode, range, and so forth, of this index can then be calculated in the usual fashion and used as a measure of the distribution of social experience among the persons in the population modeled by the graph.

Finally, a third network property which had clear implications for the analysis of interpersonal relations was the *density* or degree of completeness of the ties among elements. As we suggested earlier, the denser this pattern, the more "cliquish" the forms of interaction going on within a social group are likely to be.[35] James Coleman, Elizabeth Bott, and others had examined instances of this and, in fact, sociologists had hypothesized that a relationship existed between the "closeness" of groups and what they called *moral density* long before they developed good ways of operationalizing these concepts.[36]

Once they began using network models to describe and interpret patterns of interaction within groups, however, a number of social scientists quickly discovered a straightforward measure of the density of interconnection within a graph or subgraph. There is a finite number of ways in which a finite set of objects can be combined. This is usually expressed as the number of *combinations* of N objects taken r at a time or, simply, C_r^N. Since, in graphs, a "link" is always forged between *two* nodes, we can calculate the maximum number of connections between N nodes by substituting 2 for r in the formula for combinations and simplifying the expression

$$C_r^N = \frac{N!}{r!(N - r)!}$$

$$= \frac{N!}{2!(N - 2)!}$$

$$= \frac{(N)(N - 1)(N - 2)!}{2!(N - 2)!}$$

$$= \frac{N(N - 1)}{2}$$

Given a, the actual number of ties made in an empirical case, the density of ties within a graph or subgraph, D_t, can be then expressed as

$$D_t = \frac{a}{\frac{N(N - 1)}{2}}$$

$$= \frac{2a}{N(N - 1)}$$

"Density," in this sense, is the proportion of actual ties made as a function of the number which could have been formed. If we wish, following John Barnes, to express this number as a percentage, it can be recalculated as

$$D_t = \frac{2a}{N(N-1)} \ 100\%$$

$$= \frac{200a}{N(N-1)} \ \%$$

The two friendship networks shown in Figures 2.13a and 2.13b exemplify this. A higher proportion of all possible ties are made in the latter than in the former. This is reflected in their respective densities.

Figure 2.13a

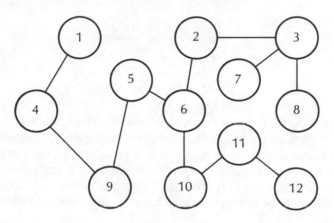

Figure 2.13b

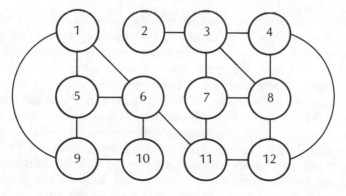

——————— = Reflexive friendship tie

If we refer to the network in Figure 2.13a as A and that in Figure 2.13b as B, then

$$D_{t_A} = \frac{2a}{N(N-1)} = \frac{200a}{N(N-1)} \%$$

$$= \frac{2(11)}{12(11)} = \frac{2200}{132} \%$$

$$= \frac{22}{132} = 16.7\%$$

$$D_{t_B} = \frac{2a}{N(N-1)} = \frac{200a}{N(N-1)} \%$$

$$= \frac{2(18)}{12(11)} = \frac{3600}{132} \%$$

$$= \frac{36}{132} = 27.3\%$$

Network A, then, is considerably less dense than network B. Simple visual inspection of these graphs, however, can be deceptive: network B is far less dense than it appears to be at first sight. By creating a formal density measure, as this example indicates, structural analysts were able to significantly improve upon the more intuitive notions of "dense" and "sparse" patterns of interaction which had been common before it was introduced.

KIN AND COMMUNITY AS SYSTEMS

While each of these properties of graphs and their associated matrices—reachability and compactness, duality, and density—had clear implications for studies of networks of interpersonal ties, social scientists, as a whole, were relatively slow to adopt them for use in empirical work. Apart from a few network *afficionados*, few researchers were even aware in the early 1960s that formal methods for analyzing patterns of this kind had developed. Conventional social science, wedded as it was to individualistic conceptions of social order and relatively simple data manipulation techniques, continued to assume that larger patterns of social interaction could best be understood by investigating the characteristics of individual actors. Even Barnes, his students, and colleagues made relatively little analytic use of systemic measures in their work.

Throughout the 1960s, however, a clear dichotomy was emerging between what we referred to in the first chapter as *aggregative* and *structural* models of social systems. By the end of the decade, the implications of this distinction had become evident to the point where what social scientists initially thought of as a disagreement over specific methods or techniques could clearly be seen as a more fundamental conflict between paradigms. As Thomas Kuhn would have it, a scientific revolution had begun.[37] One of the principal battlegrounds of this revolution was, and is, the study of interpersonal ties and communication.

Aggregative models treat individuals—or the individual elements of any system—as a collection of traits or *attributes*. These attributes are then aggregated together through the use of descriptive statistics into a general image or portrait of the population, society, or system. In some cases, subsets of elements are defined, and their characteristics form the basis for statistical comparisons. By contrast, structural models view the elements of a system as parts of a single, functioning whole. Systems, from this perspective, exhibit properties which "can best be understood in terms of the dynamic interplay among elements, i.e., the structured *relationships* among them."[38] Each element within a system thus has two aspects which are indissolubly joined: (a) a set of *unique* or specific properties which, in part, allow it to be defined as a separate entity, and (b) a set of *systemic* properties which represent the ways in which the larger whole impinges upon, constrains, and limits its latitude for independent action.

The specific interplay between a given system and its elements is, as a result, often subtle. In the radical formulation of the relationship between levels, each element reflects the essential dynamics of this interaction in microcosm. The general manifestations of a system can only be observed in their differential impact on its constituent elements. Hence, the particular or specific aspects of these units constitute the "background" against which larger, systemic effects stand out in stark relief. Without this "field," analysts would be unable to discriminate between *patterned*, or systematic, variation in behavior—which reflects larger systemic constraints—and attributes or behavior of elements which are a consequence of random or exotic factors. By contrast, in order to recognize the uniqueness of some aspect of an element, scientists must be able to factor out the effects of systemic activity, that is, to observe the deviation of an element from a *pattern*. Thus, the analytic methods used in interpreting or analyzing systems must preserve the essential unity of the processes linking the general and particular aspects of elements or risk confusing properties of one level with those of another.[39]

Social scientists following aggregative strategies of social inquiry and those adhering to the newer, structural approach described here, then, are really addressing entirely different dimensions of social process. This is not always clear to the average reader because conventional social science has adopted many of the *terms* of a "general systems" paradigm—"system" or "social system," "element," "feedback," and so on—without accepting, at the same time, either the conceptual framework or analytic consequences implied by these ideas.

The differences between these approaches, however, are immediately apparent in the ways in which those following each strategy choose to organize and represent their data. Conventional models treat systemic properties as *variables*, that is, *values* associated with each distinct element within a system which are then ordered together into arrays and manipulated for statistical purposes. Given an individual, x_1, who has three ties to other members of his work group (to borrow Bruce Kapferer's example), and given T as the variable, "tied-ness," which reflects this *individual* attribute, then the value of T_{x_1}—the x_1^{th} observation in the T array—is three. "Tied-ness" would then be treated no differently than any other individual attribute (e.g., age, sex, income) of that element; for example, it could be correlated with another variable, used as part of a regression equation, or put into a factor analysis. In effect, a system property, *connectedness*, has simply been reinterpreted in individual terms.[40]

Structural analytic methods, by contrast, seek to preserve the systemic or unitary properties of these same data by representing them in the form of graphs or matrices and then performing a series of analytic procedures *on this graph or matrix as a whole*. While some indices structuralists have developed—centrality measures, for instance, or reachability—utilize information gathered from the *perspective* of a single node or element, these data are then interpreted and standardized so as to reflect the *fit* between and among all such points within a larger system. In effect, while conventional methods disassociate observations at the level of nodes or elements from the context in which they have taken place, structural analysis seeks to retain and use this context analytically.

Kinship As a System

One of the earliest and most successful applications of truly rigorous structural analytic techniques, as we noted earlier, was to the study of kinship systems. At approximately the same time that many second-wave British anthropologists were first using rudimentary network-based techniques to interpret behavior in nontraditional societies and settings, Harrison White, a sociologist, was developing algebraic models to explain the underlying logic of the sort of closed kinship systems normally studied by his more traditional colleagues.

In his book, *An Anatomy of Kinship*,[41] White begins with the seven elementary kin roles, father (F), mother (M), brother (B), sister (Z), wife

(W), son (S), and daughter (D), defined from the point of view of ego (an abstract initial actor), which are recognized as distinct in Western kinship systems. These roles, he shows, can then be cumulated together with the role sets of others to whom ego is related to yield *trees* of compound role relations to more distant kin—for example, FZ, father's sister, FZD, father's sister's daughter. In a given kinship system, White argues, some of these compound roles may be recognized as distinct and have a specific term assigned to them, and some may not. By examining the "formal rules of cumulation" in a particular system, White maintains, it is possible to distinguish one compound role from another. Thus, for example, he argues, "Brother's father will not be a role in a society without polyandry, nor will mother's husband be a role distinct from father."[42]

From here, White goes on to develop a "minimum set of roles," that is, the smallest number of roles necessary to reflect all possible relations to ego. In order to accommodate lateral branches of families (the sons and daughters of ego's siblings, for instance), and since ego is assumed male, he adds in the husband (H) role. By compression, the mother (M) role is replaced by FW—father's wife—and daughter (D) by SZ—son's sister. This yields a minimal role set of six—*any five* of which may have direct relationships with ego at a given point in time. The structure of these roles is shown in Figure 2.14.

Figure 2.14

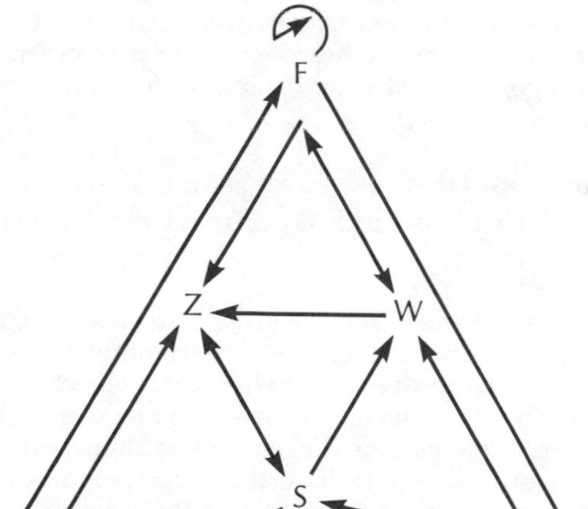

Source: Harrison C. White, *An Anatomy of Kinship* © 1963, p. 8. Reprinted by permission of Prentice-Hall, Inc., Englewood Cliffs, N.J.

In Figure 2.14 nodes are "roles" and relations are "relationships between roles." The directions given by the arrowheads indicate *possible* role compounds. Note that cyclic or repetitive role sequences are possible, for example, FWFWFW. The upwardly looping and downwardly looping arrows defined onto the father (F) and son (S) roles indicate possible paths between generations—for example, FF = ego's grandfather.[43]

Individuals would have a great deal of difficulty navigating in societies in which kinship is seen as the focal aspect of social order, White continues, unless some principle existed for creating classes of equivalent relations.

> Kin relations could be extremely confusing for ego in the small inbred society where he is trained to view everyone in a kin role, for there could be a bewilderingly large variety of roles vis-a-vis ego combined in a given person. But compound kin roles are not imposed on the society by biological constraints; in some sense they have been institutionalized "by" the society as the primary means of defining and regularizing social relations among members. One would expect that in such societies kin roles are defined and compounded according to laws that have evolved so as to minimize ambiguity in role relations among concrete persons.[44]

One way of accomplishing this simplification is to "identify" together roles which are identical from ego's perspective.

> Consider a tribe through the eyes of one man taken as ego. He wishes to place the other members at the nodes of his tree. He perceives not only role relations they have to him but also role relations between them implied by their relations to him. A new role relation, say *alter*, describes the identity relation ego perceives between two persons in the same role with respect to ego; e.g., Mac and Sam, two father's brothers, have identical roles in ego's view.[45]

Next, White stipulates a set of axioms which will allow an identification of roles of this kind to occur within a mutually consistent structure.

> Axiom IA. In the kin-role tree of a given person as ego, if two persons are alters in one node of the tree, then in any other node in which one appears the other must appear as alter to the first. . . . Axiom IB. The tree of kinship roles must be such that all persons who are alters (have equivalent roles) with respect to one person as ego must be alters with respect to any other person as ego. . . . Axiom II. If the same person can occupy two roles with respect to an ego, the content of the prescriptions of the two roles should be consistent (in the logic of that culture).[46]

White then begins using these principles to create maps of complex formal-organization-like kinship structures, called *section systems*, in which the subgroup in which members seek their wives (or husbands) is prescribed by a set of informal, but generally accepted, marriage rules.

Since in principle everyone within small, relatively bounded societies is related to everyone else at some remove, the effect of marriages is to identify roles together and hence (given axioms IA and IB) to reduce the total number of socially possible alternative roles by concretely mapping together two or more roles occupied by the same person. Each marriage, then, has consequences not only for the particular kin-relations of the individuals involved, but for the structure of the kinship system as a whole. It is precisely this *general* role structure which White is interested in uncovering.

The approach he adopts is, in one sense, quite straightforward, and, in another, quite complicated. In simplest terms, if we think of a "tree" specifying all of ego's possible kin-relations spread out in a plane, each concrete instance of the types of prescribed marriages White models would have the effect of *folding* parts of this plane together. Consequently, each "identification" between two or more roles through a marriage would have a "morphogenic effect" on the structure: it would, in effect, be created *by* instances of persons effecting marriages of the prescribed type. At the same time, the cumulative effect of these alliances would generate the structure necessary to maintain clear definitions of those subgroups needed by the system in order to perpetuate itself.[47]

After demonstrating various ways of reducing role trees, White goes on to show how particular marriage rules could lead to the creation of "closed systems" consisting of "a small number of mutually exclusive groups of alters . . . or . . . clans."[48] Next, he develops an algebraic model for describing completely integrated "section systems" based on these clans and uses it to derive a typology of distinct kinship structures under specified prescriptive rules. This model, while too complicated or, at least, too intricate to describe in detail here, has a number of salient features which are generally useful not only for understanding kinship structures, but for analyzing a range of similar phenomena as well.

First, since White is interested in kinship systems in which marriages are prescribed between particular specified subgroups, he is able to translate the constraints given by these rules into a single $n \times n$ permutation matrix[49] in which each ij referent represents a culturally specified marriage tie between "clans." Thus, for instance, if rows are taken to correspond to the "clans" of eligible men ("senders"), and columns to those of eligible women ("receivers"), each positive entry in the resulting array would show the appropriate column-group from which men in a given row-group *must* select their wives. The rest of the matrix would be zero-filled.[50]

Second, White can represent descent rules in matrix-form as well because, in section systems, all children belong to a "clan" which is uniquely determined by those from which their parents are drawn. Each positive entry in this "descent matrix," then, might correspond to the "clan" of a child (column) given that of its father (row).[51]

Third, assuming that both marriage and descent rules can be specified in this fashion, White shows how their corresponding matrices can be meaningfully combined, through standard matrix multiplication, to yield other information about a given kinship structure. Thus, for instance, simply multiplying marriage-rule and descent-rule matrices of the specific types described above will result in a permutation matrix showing the "clans" to which wives' brothers' children would belong. Other ordered powers of these types of matrices will yield products corresponding to every possible patrilateral or matrilateral clan-to-clan relationship.[52]

Finally, White demonstrates that each particular combination of marriage- and descent-rule matrices can be used to generate an algebra which will describe the structure mandated by a given set of these rules. As we mentioned in the first chapter, these kinds of algebraic relations between matrices, their extensions, or products constitute what mathematicians refer to as a *group*.[53] These groups, in turn, can be used to operationalize what sociologists and anthropologists have traditionally meant by concepts such as "structural interdependence" or "integration."[54]

Having done this, White then compares the ideal models of kinship forms, which he derives from algebraic representations of marriage and descent rules, with actual data collected by anthropologists on the use of kinship terminology in societies with closed marriage systems. The ideal models do not perfectly coincide with the actual kinship terminology. Part of the difficulty here is that "kin-term usage" and *actual* kinship are not the same thing (many of us have great-aunts or great-uncles we refer to simply as "aunt" or "uncle"). Part of it, no doubt, has to do with historic dimensions which play an important role in how actual groups come to adopt and use this terminology. In "real world" cases where the clan structure embodied in kinship terminology is well defined, however, White's formal models and observed data closely approximate one another.[55]

Thus, while White's work in *An Anatomy of Kinship* has some significant limitations—particularly in its applicability to the nonclosed kinship systems typical of modern societies[56]—it represents an important step toward a more rigorous and detailed examination of the underlying logic of kinship and its role in systems of interpersonal communication. Moreover, as we shall see in some detail in the next chapter, his formalization of kinship systems in terms of permutation matrices also led, as White suspected it might, to more powerful ways of analyzing the formal structure implicit in other sorts of institutional milieux.[57]

The Redefinition of Community

By the mid-1960s, structural analysts had largely succeeded in convincing conventional social scientists that networks of ties could be used to depict the morphology of certain kinds of social structures. The epistemological foundations of the "social network concept," however, were less well understood.[58] Thus, while second-wave British researchers—particularly J. Clyde Mitchell—had helped to establish the usefulness of network mapping *as a technique*, and Harrison White and others had shown that rigorous analytic methods could be applied to network data *under specified circumstances*, the potential range of these methods was still in doubt.

Part of the difficulty here was that most network studies conducted before 1965 had dealt with relatively small samples or bounded populations. Bott's intensive sample, for instance, included only 20 families. Mitchell and his colleagues, for the most part, confined themselves to environments in which most significant behavior could be either directly observed or collected from public sources without too much difficulty (e.g., work groups, voluntary associations, small communities). White's study of kinship, while it drew upon intensive anthropological fieldwork, was done at a high level of abstraction. The issue remained: could network-based methods be applied, in a similarly rigorous fashion, to the kind of broad-gauged and large-scale issues— urban and industrial development, occupational structure, social mobility—that had historically constituted the mainstream of much social science research? Or were network-based methods, like the sociometric studies of small groups which preceded them, destined to be confined to a relatively specialized class of problems?

The first real indication that structural analysis could address these "classic" questions began to appear in the field known as community studies (or specifically, urban studies) and involved a generalization of earlier work on "personal ties" by Bott, Mitchell, White, and others. Its central theoretical focus was the long-standing contention that urbanization and industrialization had led to an erosion of the intimate, personal, or particularistic relationships which had once formed the basis for *solidary* communities.[59]

Classical sociological thinkers had been profoundly ambivalent about the modern urban-industrial order. On the one hand, they observed that large-scale production and dense urban living had created new *opportunities* for social development: the elimination of want (Marx), the extension of universalism (Weber), the breakdown of parochialism (Durkheim), and, through urbanization, the creation of new forms of consciousness (Simmel).[60] On the other hand, it had also led to new forms of exploitation (Marx), a diminution in personal community

(Weber), new forms of social pathology (Durkheim), and a loss of identity (Simmel).[61]

During the 1930s and 1940s, however, many North American sociologists came to emphasize the second prong of these arguments almost to the total exclusion of the first. The prevailing view was, as Barry Wellman observed, that "community" had been "lost."[62] In 1959, Maurice Stein argued this case forcefully in an influential book, *The Eclipse of Community*, which summarized the results of field studies conducted in North American communities during the previous three decades.[63] Other treatments of the themes of "impersonalization" and "massification" which were written in the 1950s—such as C. Wright Mills's *White Collar* and William Kornhauser's *Politics of Mass Society*—generalized these arguments to the level of social process.[64]

Early in the 1960s, however, this orthodox view of urbanites as "disconnected," "atomized," or "massified" had been forced to confront increasingly systematic evidence—albeit on a limited scale—that personal relationships continued to play a viable role in helping individuals to find their bearings and mobilize resources within urban-industrial environments. Much of the work that succeeded in opening up these questions had been done by structuralists, and had been specifically based on the analysis of networks of interpersonal ties.

What was really at work here, of course, was a change in the definition of a "community" away from conceptions which assumed the geographic contiguousness and cultural homogeneity of the members of the community and toward ones which stressed the structure of personal bonds. By shifting focus in this way, structuralists argued, it became possible, for the first time, to look seriously at the morphology of community without explicitly accepting the social and cultural forms of the preindustrial city or village.[65]

These larger implications of the use of personal ties as an index of community structure, however, were not immediately recognized. The initial tendency, as Wellman noted, was to assimilate new, and sometimes even contradictory, evidence to old paradigms. The psychologistic models which had undergirded the traditional "mass society" perspective could be, and were, simply turned around: if urbanites were *not*, on the whole, isolated or atomized, it was argued, then the social and psychological environment of village life must have been reconstructed in some way within the metropolis. Once again, however, Herbert Gans and other researchers who argued for this position drew upon evidence largely gathered through specific, small-scale studies of bounded urban areas.[66]

Conflicting Paradigms

The stage had been set, then, for a major expansion of the debate between traditional sociologists and structural analysts over the form and substance of urban communities. The conventional view, simply stated, was that interpersonal networks had "persisted" in urban areas because they helped people to orient themselves to an otherwise untenable environment. "Community" had been *saved*, to use Wellman's term, because preindustrial social forms helped rural-urban migrants to adjust to an otherwise hostile and "alien" context.[67]

The core structuralist argument, by contrast, was that modern urban networks are *not* simply "vestiges" of earlier patterns which have been "transplanted" into an urban environment, but are the essential mechanisms through which social integration and adaptation is carried out in complex societies. Implicitly, the processes of industrialization and urbanization simultaneously created new forms of association among city-dwellers and transformed the structure of older interpersonal bonds such as kinship. Thus, structuralists argue, these urban ties, and the types of community they reflect, must be understood as phenomena in their own right.[68]

In methodological terms, this debate largely centered around units of analysis and directions of inference. Conventional sociologists, implicitly or explicitly, continued to think of network-based processes in individualistic and psychologistic terms. Networks, they argued, were simply instruments through which given individuals could affect their environment. They were, therefore, attributes of these individuals, that is, *personal* networks, whose aggregate characteristics could be used in a conventional way to describe the sets of all such individuals, and so forth.

The counterargument proposed by structural analysts was that social networks described orderly properties *of systems* and that interpersonal networks, by this token, reflected structures of ties between individuals. Therefore, they should be treated as indices of properties *of these systems*, that is, kinship systems and friendship systems, and not of individuals per se. Moreover, White had suggested that the appropriate way to ensure this was to devise methods that dealt with all relations within a system simultaneously.

While the use of personal ties as some sort of measure of "community" had become commonplace, then, there was little agreement on precisely what this meant in practice. Moreover, it was clear that resolution of some dimensions of this debate required large-scale empirical tests, because studies of small, bounded populations could be too easily distorted by differences in methods of selecting "cases." During the mid-to-late 1960s, as a result, researchers began gathering data with just these questions in mind.

Networks and Places and Bonds of Pluralism

Two studies, Claude Fischer et al.'s *Networks and Places: Social Relations in the Urban Setting* (1977), and Edward Laumann's *Bonds of*

Pluralism: The Form and Substance of Urban Social Networks (1973), clearly illustrate the differences between conventional and structuralist approaches to these issues. Both use the concept of a "social network" as a means of describing the characteristic form of communities in modern settings. Both, moreover, visualize social processes as dynamic manifestations of the patterned relationships between and among individuals and socially organized groups. And, in some measure, both utilize the same data, gathered by Laumann, Howard Schuman, and a staff of graduate assistants associated with the Detroit Area Study at the University of Michigan in 1965–66.[69] Thus, their differences do not result from slight variations in their operationalization of terms or data collection procedures, but from different theoretical and methodological thrusts.

The Detroit Area Study (DAS) data, which they both use, were assembled from interviews conducted with 985 native-born white males, between 21 and 64 years of age, who lived in the Detroit Standard Metropolitan Statistical Area (SMSA) and were selected through a multistage probability sample of "dwelling units." Each interviewee was asked to report, in detail, on his relationships with three male friends (and/or including two male relatives), their ties to one another, and their social characteristics. These raw data were then assembled into the equivalent of a series of egocentric networks of the kind studied by Bott and the second-wave British anthropologists whom we discussed earlier. Background and attitudinal data were gathered from each respondent and bare-bones background (and some relational) information was also collected from a subsample of those named by ego.[70] For the most part, however, both Laumann and Fischer confine themselves to the large-scale sample and the egocentric networks derived from it.

Fischer et al.'s *Networks and Places* seeks to explain the formation of interpersonal ties and networks in terms of what it refers to as a *choice-constraint approach:* a framework in which "individuals," by freely *choosing* among a set of "structured" alternatives, effect or create the bonds that constitute their "personal social networks."[71] They contrast these kinds of models of network behavior with ones based on "deterministic" or "mechanistic" explanations. By using a choice-constraint model, Fischer et al. contend that it is possible to discuss the origins of traditional social groupings: "Traditional corporate groups," they argue, are the "product" of restrictions or constraints on these choices. Hence, in the course of commenting on Nisbet's model of authority, they ask:

> From whence comes this authority? It derives from the power of the group over the individual. With the crumbling of corporate group walls and discovery of functional alternatives outside its confines, the power wanes, so does the authority, and so does the quality of the group. In sum, people participate in corporate groups because of limited choice.[72]

"Corporate groups," as Fischer et al. use the term, are based on "primordial attachments." They offer "families, self-enclosed villages, and traditional lineages" as examples of these.[73] When people come into contact with others outside of their corporate groups, they acquire new opportunities to form ties which cross boundaries. This, in turn, Fischer et al. argue, eventuates in less commitment on their part to corporate groups and to the norms and values associated with them.[74] "Communal ties" (those associated with intimacy, emotional depth, etc.) are thus ultimately the product of restricted choice.

> The more restricted their choice of associates, the more often and longer individuals must interact with, exchange with, and rely upon a small number of people. Thus duration, interaction frequency, and material interdependence lead to communal ties.[75]

What is going on here theoretically? Since Fischer et al.'s corporate groups include "ascribed groups" (e.g., families, ethnic groups), they appear to imply something like this: we have very limited choices with respect to the groups into which we are born. Later, however, we may choose to break these ties, and, subject to constraint, join, or join together, into new groups. Our motives in doing so are purely individualistic. Collective activity, by contrast, only comes about when we are forced, by circumstances, to cooperate.

This argument, while apparently reasonable on its face, is difficult to interpret analytically. In one sense it appears to be highly "voluntaristic." This, in fact, is how Fischer et al. characterize it.[76] However, true *voluntarism* assumes that individuals are free to select among alternatives which (a) they can mightily effect and/or (b) rationally and fully understand: it is meaningless to suggest, in any sense, that we are free to exercise our will in "selecting" the social groups into which we are born. In terms of the classic formulations of the notion of "choice under constraint," that is, game theory or decision theory, nothing is problematic here: "choices" are determined beforehand and "alternatives" are highly restricted. Moreover, one cannot simply argue that "choices" of this kind represent a limiting case: they violate the *essential preconditions of models of rational or strategic action*.[77]

Further, even granted that we exclude groups that one simply enters at birth, and granted that one's ability to choose among alternatives in forming or breaking ties is not reduced to meaninglessness by some overwhelming social constraint (e.g., prison inmates cannot simply "choose" to be free), we still face the question of whether or not people are fully able to foresee the consequences of an act like making or breaking ties. Models of rational choice assume that the risks associated with alternatives and utilities associated with given outcomes are specifiable. In this sense, the models of "bounded rationality" to which Fischer et al. refer are not qualitatively different from others: we must still be able meaningfully to assess the *degree* of uncertainty associated with both choices and outcomes. Thus, the application of the concept of "choice," in this sense, to an almost entirely ambiguous act like making a tie (selecting a friend) is inconsistent with its strict meaning.[78]

In another sense, the "choice-constraint" model proposed by Fischer et al. can be seen as highly (to use their term) "mechanistic" or "deterministic." It assumes, for instance, that a set of implied motives lies behind individuals' acts. If we can then describe the benefits which they derive from these acts in general terms, and the "tradeoffs" they are willing to make between alternatives (what economists refer to as their "utility functions"), we can, given a knowledge of the constraints, determine the aggregate choices they will make with a high degree of certainty.[79] In fact, this is precisely what economists do in *model building* and *forecasting*. Thus, the contrast which Fischer et al. suggest between models based on "choice" and "deterministic" ones is difficult to justify in practice.[80]

While precisely what Fischer et al. mean by "choice" here is unclear, their methodological rationale for using the concept is not: if each tie within a given structure may be seen as the product of a deliberate act, then any network can be fully decomposed into a series of *pairwise* relationships, counted in the usual fashion, and statistically associated with others, and so forth.[81] Throughout the rest of their analysis, this is exactly what Fischer et al. do.[82] Thus, the notion of "choice" is not a superficial or accidental feature of their framework, but the central means by which they identify their units of analysis (acts).

Apart from the assumption that "duration," "frequency," and "material interdependence" are associated with "constrained relationships," however, Fischer et al. make no systematic attempt to describe the "constraints" under which these acts occur. This is significant because most models of strategic choice assume that players take each other's moves into account. Moreover, even in the smallest networks, there are a large number of patterns which these choices may form. In the four-person "mini-networks" in the DAS data, for instance, there are 64 distinct patterns (what mathematicians refer to as "labeled graphs") which each actor must consider. In six-person networks, there are 2^{15} or 32,768. For a group the size of most seminars, say 20 people, there are approximately 1.57×10^{57}. Thus, if we assume, as Fischer et al. do by their methods, that people form each of their "individual" ties in response to those made by others, most interpersonal choices would be extremely difficult to gauge.

Historically, it was precisely these sorts of difficulties which had led structural analysts, as they became more sophisticated in their manipulation of network data, to develop a range of more powerful analytic tools. Laumann's analysis of the same data provides good examples of these.

Laumann's *Bonds of Pluralism* initially focuses on the *friendship choices* reported by his sample. First, he collapses his egocentric networks into a series of *ascriptive groups* (i.e., ethnoreligious groups), which function as *supernodes* within a larger network of ties. This allows him to be parsimonious in treating these data and, at the same time, to deal with the *social positions* of actors rather than their individual attributes per se. The ties between these "ascriptive groups" are then interpreted as measures of relative distance and mapped into the "smallest space" in which they can be adequately represented in one, two, or three dimensions.[83]

Second, he analyzes occupational groups in the same fashion (i.e., treating each occupation as a supernode, creating a distance metric between them on the basis of friendship choice, etc.).[84] Here he is principally concerned with whether "these groups form more distinctive, homogeneous occupational subcommunities with associated styles of life and value orientations . . . or that . . . actors' friendship choices are constrained . . . by the segregated ecologies of their work settings."[85]

Laumann then looks at the set of ego-centered friendship networks in terms of its homogeneity with respect to (a) those dimensions used in clustering supernodes (ethnic and religious group membership, occupation) and (b) the reported attitudes of respondents toward a range of political and social questions (e.g., political party choice, religiosity, civil liberties, work and leisure, interethnic marriage).

Finally, he discriminates among the various forms (e.g., "radial," "completely interlocking," "partially interlocking") which four-person egocentric *mini-networks* may assume, and examines the extent to which *closure* or *completeness* in them is reflected in the way in which respondents form attitudes or orientations toward issues.

Methodologically, then, this study is quite different from normal survey research. While "dwelling units" are used in sampling, and individuals (i.e., native-born white males) are selected and interviewed, Laumann and his colleagues actually employ three conceptually different kinds of units of analysis in interpreting these data: *positional actors*, created by combining sets of egocentric networks through matrix addition; *structured node sets* whose members are joined by ties, but are treated as a single entity; and *structurally discriminate mini-networks* which are used, as a class, in characterizing respondents. Each of these units has a systemic dimension built into it: positional actors are *treated* collectively, and evaluated in terms of their ability to distinguish nodes from one another within a structure.[86] Structured node sets are defined in terms of their connectivity to ego and, only then, treated as sampling units. And structurally discriminate mini-networks are taken as indices of the properties of a larger network whose parameters are unknown.[87] Thus, when Laumann begins to look at the sorts of dependent variables, for example, attitudes, which have traditionally been part of the survey researcher's stock-in-trade, he is dealing with them in a new way: as the behavioral outcome of a process which he has investigated, sociologistically, at each of several layers within a structure of interpersonal bonds.

While *Bonds of Pluralism* does deal directly with individual "acts" or "behaviors" (i.e., the expressed beliefs and attitudes of respondents), then, it does not try to explain or interpret them *individualistically*. Rather, it sees *local structure*—or the local behavior of elements—as an outcome of more general processes going on within an overarching structure of community ties. Individuals are attracted to one another, for instance, because of the mutual attractions among the groups to which they belong.[88] Similarly, homogeneity (or lack of homogeneity) in the expressed attitudes of members of friendship nets, Laumann maintains, is a consequence of the *structured* relationships among groups. Each of these types of relationships constitutes a dimension of the concrete ties between actual people. By combining these in various ways, analysts are able to construct generally higher-order models of social structure.[89]

Laumann's fundamental epistemology, however, has to be teased out of *Bonds of Pluralism:* it is not explicit, but is built into his methods. This was symptomatic of the state of the art at the time the book was written: *most* structuralists, until quite recently, have been preoccupied with developing and applying their tools. Larger explanations of their reasons for doing particular things have clearly been secondary. In some ways this is appropriate: science, in a very real sense, is "tool work." Unfortunately, however, this has probably led some people outside structural analysis to conclude that it is nothing more than a collection of tools, or "bag of tricks," which can only be applied to a specified class of problems.

While the "community debate" has never been satisfactorily resolved—due in part, no doubt, to a lack of discussion of these kinds of questions—several things are now much clearer. First, urbanites are not "massified" or "atomized": they are, as a rule, Laumann discovered, connected in a relatively dense manner into ascriptive and occupational nets. This result—which is significant in part because it rests on a randomly drawn sample and not case studies—is especially interesting because Detroit is highly industrialized. Thus, "community ties" are viable in precisely the setting where the "community lost" argument would predict they would not be.[90]

Second, Laumann's fine-grained analysis of intimate networks (which we have not presented here because of space limitations) reveals important differences in the networks of persons belonging to different ethnoreligious and occupational groups. These differences, in turn, seem to reflect structural aspects of the opportunities which members of each group have for contact with others.[91] This result, which, as we shall see, is consistent with other research, could not have been obtained, however, with either conventional aggregative methods or the relatively simple kinds of network analyses done by second-wave British anthropologists.

Finally, studies of large data sets made it clear that some reduction in the number and complexity of observed events should be an essential part of any network-based method. Thus, structural analytic techniques ought to involve some form of *homomorphic*—or "many-to-one"—data mapping at each stage of a multilevel research design. This is precisely what Laumann was able to do: to group together like classes of actors and acts, and to trace the effects of these reductions at each of several layers in a complex structure. Individualistic or psychologistic methods, by contrast, cannot be equally parsimonious without ignoring important aspects of the same forms of data.

TOWARD A TOPOLOGY OF COMMUNITY

Perhaps the most important by-product of this debate over the form of urban-industrial communities was the explicit recognition that the historic use of spatial metaphors to describe patterns of social interaction was more than just rhetoric. As Laumann and others demonstrated, "smallest space" and similar multidimensional scaling techniques are good general tools for examining problems involving "social distance." Apart from the powerful visual representations of a structure which they can provide, it was clear that these techniques can also be utilized in testing for consistency in the organization of data and their fit to various models. "Space"—or, more specifically, "social space"—was not just a metaphor, but it was still unclear how far the concept could be extended.

In broadest terms, the idea that social distance—and its generalization, social space—could be analogous to normal geometric or physical space is not new. Simmel, for instance, developed a number of ideas around spatial themes, including an explicit notion of "near" and "distant" relationships between *groups* of people.[92] Although "Simmelian space" violated some of the postulates of normal Euclidean geometry (e.g., two groups could be both "near" and "distant" at the same time), he clearly intended a close analogy between the *effects* of physical and social distance (e.g., "estrangement").[93] Robert Park, Ernest Burgess, and their colleagues and students who formed the "Chicago School" in American sociology explicitly used the term *social distance* in characterizing relations among groups. In addition, they created a research paradigm in which social distance, together with geographic, technological, and economic factors, produced an "ecology" or ecosystem which was subdivided into various "natural areas," that is, clearly delineated spatial units, each with its own specific physical, economic, and cultural characteristics.[94] Pitrim Sorokin went further in one respect and specifically suggested a close analogy between *unit-distances* in social and physical space.[95]

In all of these classic treatments of the concept, however, there was some confusion about which properties social distance or social space share with their physical counterparts, and which they do not. David McFarland and Daniel Brown (in their excellent discussion of some of these issues which appears as an appendix to Laumann's *Bonds of Pluralism*) stress the importance of *physical* "distance-like qualities" called *metrics*. A set of objects whose distances from one another may be expressed in terms of a "metric" form a *metric space*.[96]

As we shall see in detail in the next chapter, a number of techniques have been developed that allow researchers to begin with measures of similarities (between objects) which are themselves nonmetric, but which can then be used to construct essentially metric "solutions" for them. The "smallest space" technique which Laumann employed is an example of these. Some classic notions of social distance, however, suggest entirely different kinds of analogies to physical space, and would have to be interpreted differently.

One plausible explanation of why this sort of confusion occurred is that classic writers often tried to develop concepts or models which, simultaneously, incorporated measures of both physical and social space. Thus, in attempting to force a fit between the two, it was frequently necessary for them to discard properties of one domain or the other. This apparently technical decision actually implies a theoretical statement: by suppressing unique properties of one form of space, we suggest that the other is causal. As a result, when classic writers tried to force social and physical concepts of space together, they often implied theoretical connections they did not intend.[97]

Social Networks and Social Space

Among the first social scientists to empirically explore the patterned relationship between social and geographic distance were those involved in the so-called "small world" studies we mentioned earlier. In the early 1960s this group began to collect systematic, large-scale data on the *acquaintanceship volumes* of people drawn from a variety of professions and social positions.[98]

Earlier small-scale research had suggested that acquaintanceship volumes (number of "contacts" per person) varied enormously according to the social role, occupation, and so forth, of the persons involved. Systematic efforts were made to confirm these findings and to establish the broad parameters within which these loosely structured phenomena occurred. Michael Gurevitch's doctoral dissertation, "The Social Structure of Acquaintanceship Networks," is probably the best example of work of this genre.[99]

In the late 1960s, Stanley Milgram first reported the results of experiments, utilizing what he called "small world" techniques, which were designed to trace out the *contact chains* (i.e., indirect transitive linkages) of people drawn from a larger population. Initially, a group of *starters* was selected. Each starter was given a packet of instructions and the name of a *target*. He or she was then asked to pass the packet along to that target according to rules which stipulated criteria which had to be met before a transfer could occur. For instance, in a given experiment a sender might be required to send the packet only to people he or she knew on a first-name basis. If a given starter did not have this sort of relationship with the target, he or she was instructed to pass the packet to someone who might. This process was repeated until the packet arrived at its intended destination and the experimenter was contacted. At each stage, postcards were sent back to the project office so that incomplete chains could be traced. In the initial trials, starters and chain members were given a range of data about the target (e.g., address, occupation) and were asked to answer questions about their own characteristics. Later trials, which continue up until the present, systematically vary the forms and kinds of identifying information given to subjects.[100]

Among other things, these ingenious experiments reveal a number of important features of the organization of social space. For instance, a study conducted by Jeffrey Travers and Stanley Milgram in 1970 shows that the social distance between senders and receivers (as measured by the number of passes of the packet) may be smaller within occupational-professional chains than it would be in more geographically bound networks, such as ones based on "residential location." Thus, in this experiment, it took senders an average of 4.6 links to reach their target through business contacts and 6.1 links through residential channels.[101] This result and those obtained by other researchers suggest that the chain-length of occupational-professional networks may *not* vary linearly with geographic distance. Thus, if we were to plot an individual's "social neighborhood" by placing all individuals who are the same social distance from ego at the same grid coordinates away from him or her, the persons at each of these points would not necessarily be *socially* close to *one another*. This property of social space violates our presuppositions about the geographic contiguousness or *geolocality* of "neighborhoods."[102]

Similarly, Charles Korte and Stanley Milgram conducted a "small world" experiment in which 540 people were asked to begin chains toward 18 male targets in New York City—nine of whom were white, and nine black.[103] Target groups were matched with respect to "age, income, education, occupational status and number of organizational memberships."[104] Thirty chains were directed at each target. Starters were recruited in Los Angeles through a mail solicitation procedure. The race of the target was not included in the packets used in the experiment, although someone with a good knowledge of New York City might have been able to guess the target's race from his street address.[105]

The results of this study strongly indicate that there are a number of structurally significant differences between the organization of white and black contact networks. First, more white-target chains were completed than black-target chains (33 percent opposed to 13 percent). Second, completed black-target chains were longer (5.9) than completed white-target chains (5.5). Third, at any point in a chain, the probabilities of a successful pass of the packet were higher between whites than between whites and blacks, or between blacks. Fourth, convergence effects (a common chain-member in more than one chain) occurred with respect to 9 of the 18 targets: 7 white targets and 2 black. Finally, as a rule, packets primarily passed through white networks had a higher probability of success in reaching black targets than those routed exclusively through black networks.[106]

If these observations are correct, they suggest that there may be "breaks" or discontinuities in some groups' social spaces. While the reasons why Korte and Milgram obtained their particular results are unclear, the fact remains that *lack of continuity* in a group's social space may indicate that its members have a great deal of difficulty in mobilizing resources, exchanging information, or fomulating common goals and strategies.[107] Moreover, if due to these discontinuities, one group is effectively dependent upon others to "carry" socially necessary communications, the dependent group may be effectively barred from undertaking certain types of independent action to protect or enhance its interests.

Social Space and Community

Spatial representations of interpersonal networks, then, may ultimately provide us with answers to a number of very subtle and otherwise intractable questions about the nature of "community." Recent studies, such as those being undertaken by Nan Lin and some of his colleagues,[108] have begun to refine experimental designs so that the "communications" process modeled by the small world technique can be related to some of these larger issues. As we shall see in Chapter 4, Laumann and some of his colleagues have made extensive use of spatial models in examining the structure of influence and power in communities.

Barry Wellman's article, "The Community Question: The Intimate Networks of East Yorkers," however, provides a clue as to what may become the most important application of spatial models in the study of community structure. The branch of mathematics that deals with the properties of surfaces in many dimensions is called *topology*. In 1972, René Thom, a French mathematician, published a book entitled *Stabilité Structurelle et Morphogénèse* (*Structural Stability and Morphogenesis*) in which he describes the properties of a set of topologies, called *catastrophes*, which may be used to model sudden or "unpredictable" changes in behavior.[109] While the specific illustrations of *catastrophe theory* which Thom provides are still controversial, his general message to social scientists is clear: structural analysts and others interested in the morphology of social structures must consider the *structural transformations* which may result from gradual changes in the social spaces they have been examining.[110]

Wellman and other sociologists who have been connected with the Community Ties and Support Systems Project at the University of Toronto strongly suggest that this is exactly what may have happened in the course of the development of modern urban communities. Since 1968, they have been studying the primary ties of people living in the borough of East York, "an upper-working-class/lower-middle-class, predominantly British-Canadian, inner suburb." East York is typical of many suburban areas in North America: most of its residents live in "small private homes or high-rise apartments," with rarely more than two adults per household.[111] It is precisely these kinds of areas which social critics had in mind when they argued that "community" had been "lost."

What Wellman and his colleagues found out, as we might suspect by this point, was that East Yorkers are involved in a plethora of "intimate" ties which they use in order to garner resources, deal with crises, and cope with bureaucracies. These ties, however, are not primarily with neighbors or other people living in East York, but extend throughout Metropolitan Toronto and beyond.[112] Thus, while East Yorkers are clearly involved in a "community," their *social neighborhoods* do not directly coincide with the physical areas in which they live. Wellman goes on to argue that this transformation in the spatial organization of urban ties, together with an apparent differentiation in the functions of different kinds of interpersonal bonds, has "liberated" modern forms of "community" from the strictures of geographic proximity.

SUMMARY

The models and techniques which structural analysts have been developing during the last two decades have enabled them to begin examining personal dimensions of social life without dwelling on the conscious and unconscious motivations or individualized behavior of actors. In the course of creating and employing these models, structuralists have elaborated an alternative to the conventional psychologistic view of society as a simple product of individual encounters, reciprocal orientations, and interpersonal alliances.

The conceptual basis of this new framework is the notion that kin and community may be thought of as systems whose organization can best be modeled by structures of interpersonal bonds or ties. These structures, in turn, have properties—reachability and compactness, duality, and density—which are formally analogous to features of the "real world" they are intended to represent. In contrast to traditional aggregative approaches, structural models view the elements of a system as part of a single functioning whole which can be best understood in terms of the dynamic interplay between its "general" and "particular" aspects. In methodological terms, this perspective demands that analytic operations be performed on the graph or matrix representing a structure as a whole, as well as on each of its component parts.

White was able to construct structural models of this kind for closed kinship systems with prescribed marriage rules. By creating a minimum set of roles which can be compounded together to yield more complex kin-relations, he was able to generate an a priori map of kinship relations as seen from the perspective of ego. This map can then be structured by applying a set of axioms to it which reflect alternative marriage prescriptions. By extension, White was able to devise an algebra which represents the logical consequences of prescribed marriage rules in societies based on "section systems." While this model is limited in its applications to modern urban kinship systems, it suggests powerful ways of analyzing the formal structure implicit in other kinds of settings.

Beginning in the mid-1960s, structural analysts were able to develop a range of techniques which used interpersonal bonds between individuals as an index of the structure of urban-industrial communities. By redefining the concept of "community" in this way, they were able to formulate hypotheses about the form and substance of contemporary city life which were not implicitly bound to the social and cultural morphology of preindustrial society.

By this point, however, many conventional sociologists had begun to use network-like data, and terms derived from network models, as a means of describing the "attributes" or aggregate characteristics of individuals rather than the systemic properties of patterns of kin or community ties. A close comparison between Fischer et al.'s *Networks and Places* and Laumann's *Bonds of Pluralism* shows, however, that conventional treatments of network data ultimately rest on voluntaristic and individualistic assumptions, which are both theoretically and technically difficult to justify in practice. By contrast, the more powerful methods and tools developed by structural analysts are able to treat these same data parsimoniously by homomorphically reducing networks into like classes of "actors" and "acts" at each of several layers within a complex structure.

Two of the specific techniques structuralists have developed—"smallest space" mapping and "small world" chains—strongly suggest that the study of social space per se may yield important insights into the structural transformations which have taken place in the form and substance of community in modern urban-industrial society.

In the chapter which follows, we will examine the consequences of applying structural analytic methods to the types of social and economic systems which, many argue, were responsible for the creation of the modern urban and industrial order—and, by implication, the patterns of community structure which we have focused on here.

Note: The exercises here, and at the end of Chapters 3 through 5, are intended to provide readers with an opportunity to explore the principles or techniques of structural analysis discussed in the text. Each exercise involves some minor data gathering and/or analysis. Computers may be used in performing these analyses, but they are not essential: readers will find that their intuition and some elementary measurement are generally more useful than an elaborate software package. If these exercises are being completed as part of a class assignment, students will find it useful to work on them together.

EXERCISES

1. Construct a table showing the kinship relations among a set of people. (If you do not have a convenient group from which to gather data, your library will have published genealogies.) Be sure to specify the "mapping rules" you use in designating nodes and relations. Are the nodes "individuals"? "nuclear families"? persons living within the same "household"? Are the relations between people "sharing a close kinship bond"? "lines of descent"? Do "sibling ties" count? You may want to construct several different types of ties defined onto the same nodes, and then compare them.

   After you have constructed this table, draw one or more networks of kinship ties—both nodes and relations—in graphic form. Taking powers of t from some arbitrary node, make a list of the kinship terms we use to designate the relationship of nodes at this distance from ego, e.g., "first cousin," "uncle." Then, vary *one* of your mapping rules, and repeat the process. Does this change make a difference? If so, how? If not, why not?

2. Construct a matrix showing one type of affective tie defined onto the members of a small group, taken as nodes. (Once again, if no group is handy, use one of those given in the sources cited in the notes to this chapter.) Using the rules for matrix multiplication specified in Definition 3, square and then cube this matrix. Draw graphs corresponding to each of these products. What do they show? Write a paragraph describing the results of each of these operations.

Construct a matrix corresponding to another set of affective ties defined onto these same nodes. Given that your original matrix is designated as A and your second as B, show the results of each of the following operations: $A \times B$; $A + B$; $A - B$; and $A^2 + B^2$. Present a brief interpretation of the social patterns to which each of these operations corresponds.

3. Choose a small part of a "community" (e.g., part of a village, church or ethnic group, neighborhood). Define a set of ties, *other than kinship*, onto the members of that group. Collect from the members of that part of the community the terms they use to designate their relationships to one another in terms of the type of tie you are examining. Using the method described in White's *Anatomy of Kinship*, and referred to here, create a set of roles vis-à-vis some ego which adequately and consistently describe all ties. Then, following White's procedure, see whether this set can be reduced to a minimal set. In what ways are your sets of roles similar to and different from White's?

ADDITIONAL SOURCES

Barnes, John A. *Social Networks*. Reading, Mass.: Addison-Wesley, 1972.

Boissevain, Jeremy F., and J. Clyde Mitchell (eds.). *Network Analysis: Studies in Human Interaction*. Mouton: The Hague, 1973.

Cohen, Abner. *Custom and Politics in Urban Africa*. Berkeley: University of California Press, 1969.

Katz, Pearl. "Acculturation and Social Networks of American Immigrants in Israel." Ph.D. dissertation, SUNY Buffalo, 1974.

3

Corporations and Privilege: Economic Structure and Elite Integration

THE SETTING

Few social inventions have had as profound an impact on the quality of modern life as the private business corporation. As late as the beginning of the nineteenth century, incorporated bodies held only a tiny proportion of the capital assets in most countries. Stock companies, where they existed, were small relative to the wealth of their principal stockholders or directors and functioned, for the most part, as extended forms of partnership. Few companies, incorporated or otherwise, had more than 100 employees.[1]

Today, no more than a few hundred of the largest corporations in Western capitalist economies typically control between 60 and 65 percent of all productive wealth and employ 40 to 60 percent of all those working in the manufacturing or mining industries.[2] No individual—not even one of their largest stockholders—is as wealthy as the largest *multinationals*.[3] No university has as many scientists, engineers, and technicians associated with it as one of the major corporations. Few law firms have as many lawyers.[4] Only the very largest national governments, in fact, can rival the leading corporations either in the variety of activities in which they engage or in the depth of expertise they can bring to bear on problems.[5]

In some cases—particularly in Japan—these private businesses act, in certain respects, like socialist states, providing housing, retail stores, health services, day care, banking services, education, insurance, food, and recreation for their employees.[6] A few analysts have even suggested that given their worldwide scope and profound impact on the political, social, and cultural institutions of the countries in which they operate, the largest private business corporations collectively constitute the only truly effective form of supranational organization.

Despite this, we know relatively little about many aspects of corporate organization, behavior, and structure. Comparatively few detailed studies of specific corporations have been reported in the open literature.[7] Most case studies examine highly circumscribed aspects of corporate performance: accounting practices, advertising and marketing, hiring and promotion, and so on.[8] Researchers who focus on more sensitive issues typically cannot gain access to inside information and are thus forced to rely upon statistics derived from public sources (e.g., annual reports, "insiders" transactions, stock registrations, press releases) which have been culled by business information services and assembled into directories.[9] As a result, their data seldom, if ever, reflect uniform reporting standards and hence are often inaccurate in detail.[10]

More broadly gauged studies—ones which attempt to deal with the behavior of business concerns within industries and markets—are almost always confined to *aggregate* characteristics of firms (e.g., their assets, relative to the total assets of sellers or buyers; their sales, compared to other sellers; their shipments, relative to those of other businesses operating in the market). Similarly, the effects of corporations on their "environment"—on the people who buy goods and services, on the communities in which they carry on business, on the distribution of wealth, and so forth—are almost always also treated aggregatively. "Corporate behavior," in other words, is traditionally thought of as a species of *individual* behavior in which the "individuals" involved are business concerns rather than people. Modern industrial corporations—whether they are seen as physical embodiments of Adam Smith's "Economic Man" or, less anthropomorphically, as "rational optimizers"—are invested by conventional model builders with a simple vocabulary of motives and a limited repertoire of action. The "markets" in which they operate are, for all practical purposes, unstructured.[11] Hence, apart from general "market forces" such as supply and demand, the only effective constraints on individual firms' actions which conventional analysts willingly acknowledge are those brought about by the activities of other firms within the same product markets.

Structural analysts contend that this individualistic and psychologistic approach to modeling economic systems, while often consistent and parsimonious, does not adequately reflect the complexity and interrelatedness of modern economies. What is usually missing, they contend, is some explicit and formal recognition of an overarching structure which constrains the behavior of individual economic actors and orchestrates their relationships to markets, to industries, and to one another.

In this chapter we will first examine a number of structuralist models of economic systems, per se, and a series of attempts to use these to explain the observed behavior of actors in a variety of "real world" settings. We will then turn to an important and, as yet, unresolved issue: the relationship between economic systems and structures of privilege or class.

Conventional and Structural Models of Economic Units

Economic models traditionally rest on two key concepts: the idea of a *firm*, or private business concern, and the notion of a *market*. In practice neither is well defined.[12] Firms, for instance, are sometimes delineated on the basis of their functions within an economy, industry or market and, sometimes, their legal form or status.[13] In the same vein, markets may be treated either as "constructs" or as tangible things ("concreta"). Scholarly articles will often use the same terms in more than one analytically distinct way, and, as a rule, pay little attention to the epistemological issues involved.

This ambiguity is, to some extent, deliberate. The prevailing "Structure-Conduct-Performance" paradigm is flexible enough to accommodate a number of qualitatively different interpretations of economic *behavior*, and, at the same time, sufficiently restrictive to preclude serious attention to system morphology or structure.[14] In effect, it retains surface consistency by tolerating contradictory definitions of units of analysis. By implication, serious attempts to resolve these disagreements could force a reconsideration of some fundamental aspects of the paradigm itself.

Since the early 1960s, this is precisely what a number of structural analysts have been trying to do. Two distinct lines of attack have been followed. The first, which evolved out of *exchange theory* and studies of complex organization, emphasizes the extent to which routine processes of exchange or interaction tend to produce "dependency" and, hence, "asymmetric" relationships between the elements of a system.[15] Ordinary transactions, it argues, would normally lead one or more firms to attempt to dominate or control others through mechanisms that were only indirectly related to the forces of supply and demand in the marketplace. *Inequality* in the exchanges between specific market participants, would, in turn, generate a systemic pattern in which nominally independent or autonomous economic actors would form into a series of hierarchically arranged cliques or clusters of elements functioning, for all practical purposes, in concert with one another. As a result, structuralists doing research along these lines have been systematically exploring nonmarket mechanisms, such as outright ownership of voting stock or control over appointments to boards of directors, and their implications for the market conduct or performance of corporations.[16]

The second line of attack ultimately derives from the work of

Rudolph Hilferding, the Austrian Marxist economist. In a famous treatise, *Das Finanzkapital (Finance Capital)*, first published in 1910, Hilferding observes that commercial and industrial corporations in advanced capitalist countries increasingly tend to be dominated by commercial banks, investment syndicates, and other financial or fiduciary institutions.[17] As joint stock companies expand, Hilferding argues, they routinely depend upon capital markets for short-term credit and for fresh supplies of long-term investment capital. Banks and other financial intermediaries, because of their central role as underwriters of stock issues and as agents in the issuing and redemption of bonds, are able to gain commanding blocs of shares within individual corporations by simply retaining them for their own portfolios. Given the dispersal of ownership within a corporation which normally accompanies the expansion of its capital base, small, concentrated stockholdings, such as those acquired by banks in this fashion, convey more effective power over decision making than larger but more amorphous holdings of the majority. As a result, structuralists pursuing this line of reasoning have focused on the central role of financial institutions within networks of corporate ties and have emphasized the primacy of bank-owned shareholdings and bank-appointed directors within corporate structures.[18]

While they originate in two entirely different theoretical traditions, these structuralist interpretations of the dynamics of corporate systems contrast sharply with conventional approaches to economic questions. First, both emphasize the importance of intercorporate ties and connections in shaping or determining the conduct of individual firms within industries and markets. Conventional economic models simply assume that the units with which they deal—firms, buyers, corporations, or whatever—are distinct. Structuralists argue that the degree of independence of *nominally* separate economic units must be treated as problematic in each case, and that, as a result, the intricate web of interconnection between firms in modern economies is a theoretically and practically important fact which must be explained.

Second, structuralists are aware that complex systems are multi-tiered and that, consequently, important types of interaction may be going on at each of several layers within them at the same time. Conventional models of economic systems are concerned with only one of these, namely, exchanges between "buyers" and "sellers" within markets. Structuralists contend that other forms of interaction may bear on market conduct in important ways and, hence, that models should be constructed in such a way that these patterns may be taken into account.[19]

Third, structuralists distinguish between general effects which take place, and hence must be observed, at the level of a system as a whole, and particular effects which may be localized within a specific element or set of elements. Thus, it is extremely important, they argue, to clearly define both the units of analysis used in modeling a structure and the operational distinctions among them. Most economists only deal with businesses or corporations as isolated actors (e.g., buyers and sellers within given markets or industries). This effectively precludes an examination of the context in which economic transactions occur. Hence, the information needed to draw fine distinctions between units of analysis is often lost. Structural models, by contrast, preserve this context in order to detect the degree of transitivity in the effects of one part of a system on another.

Fourth, structuralists recognize that models are only analogs to a more complex "reality." Concepts, such as the notion of a "market," are thus only "constructs," that is, abstractions of "real-world" processes or things. Hence, one cannot refer to them as if they were tangible. Conventional economists, however, often attribute particular empirical outcomes to the "operation of the market" or of "market forces." Structuralists maintain that theories (or models) and data may be juxtaposed to and compared with one another, but that one cannot logically demonstrate or "prove" an empirical statement by referring to a theoretical one.

Finally, both of these structuralist approaches emphatically reject the idea that the behavior of firms or corporations can be explained in terms of a simple vocabulary of motives. Corporations are not, structural analysts argue, simply *persona ficta*, or legal persons, which behave in all important respects like individual human actors. Corporations are complex organizations whose internal and external structure interacts in complicated and often dynamic ways. Moreover, the actions or behavior of one corporate entity is contingent upon the actions or behavior of others. As we saw in the previous chapter, models of strategic decision making that assume perfect independence quickly become unmanageable. Thus, structuralists contend, models of corporate behavior or structure must systematically reduce this complex "reality," but without confusing different types of system effects with one another.

Directorship Interlocks and Corporate Structure

Given this emphasis on intercorporate connections and/or exchanges, structuralists could easily have used a number of different types of empirical evidence to test or examine theories about corporate systems. However, while later studies frequently employed data on shareholding, most early structural analytic research concentrated on describing or analyzing *interlocking directorates*, that is, "bonded-ties" which come about when two or more corporations share a member of their boards of directors. In retrospect, there appear to have been two reasons why structuralists initially focused on these particular types of ties.

First, interlocking directorates occur in all modern capitalist economies.[20] Typically, they form extremely complex patterns, and even small samples will yield several hundred firms which are linked at t^1 and t^2 to the largest banks or industrial corporations. In most cases, networks of directorship interlocks are quite dense. This density, moreover, does not appear to have decreased substantially since the turn of the century—in the face of substantial increases in the number of corporations involved. In spite of their ubiquitousness and persistence, however, no adequate theory existed to explain interlocking directorates. Mainstream economics either ignored them entirely or argued, without empirical evidence, that they were meaningless. Unconventional social scientists—who actually investigated them—found that the simple *number* of interlocks between firms was not strongly associated with most measures of corporate performance.[21] As a result, structuralists felt that directorship interlocking constituted a rich vein of largely unexplored data which, because it was inherently relational, might be amenable to precisely the sorts of tools they had been developing.

Second, from a purely technical point of view, directorship interlocks were, superficially at least, more reminiscent of the types of person-ties that other structuralists had been studying than the kinds of measures that they could have devised from information on the plethora of stocks, bonds, borrowings, credit lines, and so forth, with which modern corporations structure their relationships to one another and to capital markets. Thus, even those analysts who saw themselves as following in Hilferding's footsteps were reluctant to delve into other forms of data without first dealing with a phenomenon which was similar to ones which were reasonably well understood at the time.

By the mid-1970s, structuralists had undertaken a number of important studies of interlocking directorships.[22] While this research reflected many of the same general theoretical orientations and substantive interests as work previously done by traditionally trained social scientists, it differed from it in a number of important respects. Many conventional researchers studying interlocking directorates, for example, had used *men*, rather than *institutions*, as units of analysis.[23]

Before the late 1960s, almost all directorship data were simply counted, cross-tabulated (by man), and presented in the form of a distribution—whatever the particular analytic purposes a given researcher might have had in mind. In most cases this reflected an implicit interpretation of directorships as personal communications channels. C. Wright Mills's famous critical study, *The Power Elite* (1956), for instance, emphasizes the importance of interlocking directorates in knitting together a "community of interest" by providing channels which ". . . permit an interchange of views in a convenient and more or less formal way among those who share the interests of the corporate rich."[24] For Mills, in other words, interlocking directorates are important because of their consequences for "the power of individual men," not principally because of their institutional effects.

> In fact [Mills argues] if there were not such overlapping directorships, we should suspect the existence of less formal, although quite adequate, channels of contact. For the statistics of interlocking directorates do not form a clean index to the unity of the corporate world or the co-ordination of its policy: there can be and there is co-ordinated policy without interlocking directors, as well as interlocking directors without co-ordinated policy.[25]

The monograph that Mills cites as background for this discussion, a U.S. Federal Trade Commission study completed in 1951, is symptomatically concerned with the fact that approximately 1,500 of the 10,000 or so directors associated with the 1,600 leading U.S. corporations held seats on more than one board.[26] Its treatment of the institutional effects of these ties, however, is largely hypothetical, and bears little direct relationship to its data.[27]

This is typical of much of the research of this genre. Since they implicitly focus on a network of men (nodes) linked by corporations (relations), the inferences these studies draw about corporate structure are often weak or sketchy. Sam Aaronovitch's *The Ruling Class: A Study of British Finance Capital* (1961), for instance, seeks to establish the dominance of commercial or merchant bankers by showing that they frequently appear as members of the boards of a wide variety of industrial corporations, joint-stock (or public) banks, insurance companies, and so forth. On the basis of these data, he goes on to distinguish 31 "groups" of firms which usually "center" around various commercial or merchant banks or investment houses and are, presumably, dominated by them.[28] Michael Barratt Brown, in a study aptly entitled "The Controllers of British Industry" (1968), applied much the same sort of analysis to later British data and obtained results which are consistent with Aaronovitch's.[29] Similar research has also been conducted on "corporate elites" in Holland, Germany, Canada, and other countries.

The chief difficulty with this approach, of course, is—as we noted in the last chapter—that there is no necessary correspondence between a given network, its isomorphs, and their possible duals. Thus, resting any inferences about overall corporate structure on an assessment of man-based networks is, at best, risky. For instance, even if we assume that all directorships are equally important, there is no guarantee that

men who are "central" to the network of man-to-man relationships will
be associated with corporations which are "central" to the corporate
structure. Figures 3.1 and 3.2, and 3.3 and 3.4 illustrate some of the
problems here.

Figure 3.1

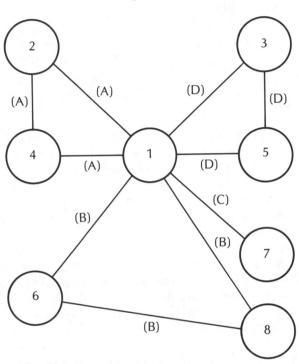

In Figure 3.1 a series of men, designated by numbers, are tied
together by their memberships on the boards of various companies
(indicated by letters). Note that man 1 is most central. In fact, he would
clearly score highest on the basis of either of the centrality measures we
mentioned in Chapter 1. One of the reasons why this is the case is
apparent in Figure 3.2: man 1 forms the only interlocking directorate in
the corporation-to-corporation mapping. No corporation in Figure 3.2,
however, is any more or less central than the others; each is connected
into the net equally tightly. In this case, if we had identified man 1 as
"central," and used him as a "tracer" or pointer, we would have gained
no information about the relative centrality of firms.

Figure 3.2

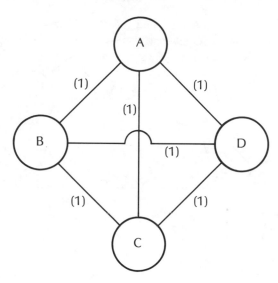

Figures 3.3 and 3.4, by contrast, provide one instance where focusing on the number of directorships held by one given man would not merely be distracting, but could be seriously misleading. Note that the network shown in Figure 3.3 consists of one densely interconnected clique and one hanger-on. If we represented these data in tabular form—as most conventional social scientists looking at interlocking directorates in this way usually did—each of the men in this clique would be credited with five $(N - 1)$ directorships. Man 7, however, would appear to be relatively unimportant. If we had excluded him from further consideration on this basis, however, we would have missed the extremely interesting and fairly central role played by corporation G, which is shown by the dual of this network in Figure 3.4.

By the late 1960s most, but not all,[30] sociologists working in the area had come to recognize that man-to-man networks tell us very little about corporate structure, per se. However, they continued to use simple cross-tabulations as the principal means of representing their data, simply substituting "firms" for "men" as their units of analysis. As late as 1967, Lloyd Warner and Darab Unwalla, pursuing a model derived from the exchange theory/social organization literature, employed simple classifications of *types* of directorships as a means of describing what they explicitly referred to as "the system of interlocking directorates."[31] In 1969, Peter Dooley used more sophisticated methods

Figure 3.3

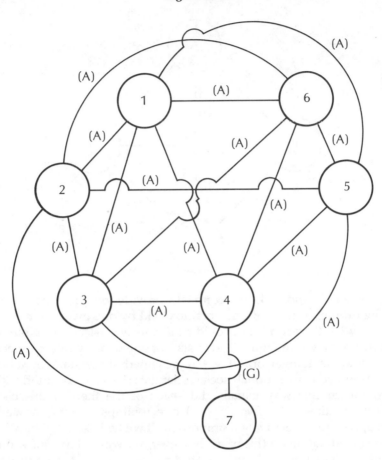

and a more detailed typology of corporate types to examine essentially the same sort of data, namely, a "top-down" sample of the ties among the largest U.S. firms. While Warner and Unwalla had *snowballed* their initial sample, however, Dooley had not. Despite this, his substantive results were quite consistent with theirs.[32]

While Dooley's formal approach was essentially aggregative, the logic underlying his analysis was clearly structural, and he made several attempts to introduce structural and quasi-structural methods into it—including a rough-and-ready measure of tie-density.[33] In this sense, his study was probably the clearest ancestor of the work which structuralists were to undertake during the decade which followed. The beginning point for their own development of a distinct analytic perspective on interlocking directorates, however, was the recognition of an essential error in earlier studies of networks of firm-to-firm ties.

Figure 3.4

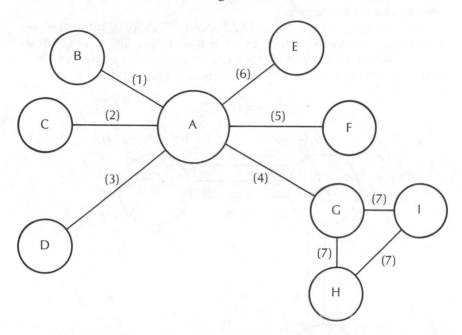

In the course of shifting their focus away from networks of men linked by corporations to networks of firms linked by men, conventional social scientists had not taken into account a fundamental property of nodes which had been changed in the process. In a man-to-man network, the unit values of nodes are essentially fixed. Corporate boards, however, can vary a great deal in size. Thus, the relative size of corporate boards acts as a constraint on the ability of firms to interlock with one another.

For instance, given two corporate boards of different sizes, A and B, if the size of A, O_A, is greater than that of B, O_B, then the *maximum* number of directorship ties between them is equal to O_B. Formally, Max $(N) = $ Min (O_A, O_B) where N is the number of shared ties between them.

Even in the case of pairwise connections between firms this effect is complex. Note, for instance, that this size limitation acts as a constraint on *both* senders and receivers: A cannot send ties to B which B cannot "receive." Thus, even if A and B are essentially the same firm operating under two different legal names—and A, as a result, completely encompasses B—an *attribute* measure taken on A would be limited by the relatively arbitrary size of B's board.[34] Hence, a large firm associated with a series of completely dominated smaller firms would appear to be less important than a similar firm which was tied into the same number of smaller firms, but with arbitrarily larger boards. Where three or more firms share ties, directly or indirectly, the potential effects of these size constraints are even greater.

In Figure 3.5, three firms (A, B, and C) share a total of seven interlocking directorships: four join A to B and three link A and C. If we used the simple number of such ties as an adjacency measure, A and B would be reckoned as "closer" than A and C or B and C.

Figure 3.5

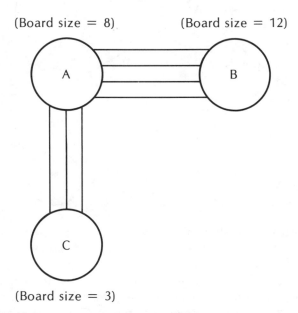

(Board size = 8)   (Board size = 12)

A   B

C

(Board size = 3)

Source: Reprinted, with permission, from S. D. Berkowitz, "Structural and Non-structural Models of Elites," p. 20.

Calculating adjacency in this fashion, however, would ignore important features of these data. Note that the entire board of C is, in effect, a subset of A's board. Given no further information, we might hypothesize that a group of directors on A were either using C as a vehicle for their own private investments or that, simply, C is some sort of subsidiary or partner-firm of A. The relationship between A and B is more ambiguous. While the overlap is high (.5 of A and .33 of B), given the information at hand we cannot determine the relative influence of one board, all things being equal, on the other. We do know, however, that the directors A shares with B are *not* the same as those it shares with C: if this had been the case, there would have been direct ties between B and C in the graph.

Thus, while the *proportion* of ties shared by sending and receiving nodes gives us a finer picture of what is going on between them, it will not solve all our problems. Apart from enumerating the sending and receiving proportions of each pairwise connection in some larger graph, it is not readily apparent how one could go about describing these proportional effects where the number of nodes in a network is large. Moreover, the most significant datum represented in the graph may very well be the *absence* of any direct connection between *B* and *C*. Here, of course, proportions are no help at all.

STRUCTURAL MODELING OF INTERCORPORATE TIES

Before going on to a discussion of the theoretical and substantive impact of structuralist studies of directorship interlocks and other forms of corporate interconnection, let us briefly consider some of the purely methodological solutions that structural analysts have devised for problems of this kind. In previous chapters we mentioned two powerful techniques that structuralists have created or adapted for dealing with complex structures: multidimensional scaling and algebraic modeling. Both have been successfully applied to problems of corporate or organizational structure. Although some of the more technical aspects of how structuralists implement these techniques are beyond the scope of what we could hope to cover in a volume of this kind, the basic principles involved are relatively straightforward and essential to an understanding of the contributions that structural analysts have made to the description and analysis of corporate systems.

Scaling Techniques: The Basic Ideas

In 1972, Joel Levine published a study, "The Sphere of Influence," which not only provided a clear model for the application of structuralist techniques to complex system problems, but which soon established itself as a standard or benchmark for later research into corporate structure.[35]

Levine's task was to represent, in the most parsimonious and consistent way possible, a large number of ties between banks and industrial corporations in the United States (1966): 84 nodes and 150 links.[36] His theoretical goal, therefore, was *not* to establish the set of underlying "laws" or lawlike statements which governed the creation of these ties (what we call a *nomothetic* model), but usefully to describe the configuration or pattern of interconnection—to create an *idiographic* representation.[37]

To be useful later on—when we are trying to "make sense" out of what we have done—this sort of map must be relatively simple. If it is as complicated as the original data, there is little reason to construct it in the first place. If it is too simple, by contrast, we will find that, just like a poor road atlas, it is likely to lack those minimal details needed in getting from one point to another. The happy medium is to find a "mapping" or guide that preserves those essential dimensions of "reality" needed for navigating.[38]

Levine's problem is directly analogous to this. If the pattern of linkage between a small number of nodes (corporations) is quite simple—as, for instance, in Bott's or Young and Willmott's networks—then an ordinary line drawing will do. No more complicated "picture" is needed. If we wish to incorporate some measure of "closeness" into it, this can be done by separating the nodes by a scale-distance which reflects some sort of proximity measure.

This is possible, but tricky, because we have to find a mapping that will satisfy all these distances *simultaneously*. For a network of N points, we can create a "perfect" or exact representation in $N - 1$ dimensions. Thus, where a network contains one node (and, of course, no relations) this can best be shown in zero dimensions.[39] A network where $N = 2$ can be exactly represented in *one* dimension—in effect, as a straight line. For values of N greater than four, while interpoint distances can be calculated, it is difficult to represent them in a space which is intelligible to most people. Thus, in order to create an adequate map where $N \geq 4$ it is necessary to *compress* these distances in such a way as to yield a more-or-less accurate three-dimensional picture. In most cases this can be accomplished reasonably well by employing techniques generalized from those used to find relative distances in "normal" or Euclidean spaces.[40]

Levine begins with a list of 14 banks and 70 industrial firms which form a single, completely connected component of a larger graph.[41] The shortest-path distances between nodes are also small: eight of the banks are separated from one another by only one intermediary. The next-shortest paths between banks are at t^4; and all six remaining banks are this distant from one another. Thus, the network as a whole is highly compact, and there are a large number of, in effect, redundant indirect ties between given pairs of nodes.

Before Levine can create a map of these data, he must first establish some kind of rough *quasi-metric* measure of the distances between points.[42] In order to do this, he separates his nodes into two subsets—banks and industrials—and uses bank-to-bank connections as an indirect *similarity* measure for the industrials. He then repeats the procedure using corporations as "judges" of interbank distances. These two sets of measures are then combined, using a technique called *unfolding analysis*, to yield one in which *the more directorships* two nodes share the more "similar" they ought to appear to the members of the other set ("judges") and, therefore, the closer they ought to be mapped to one another. By reducing the dimensions of this measure, he is then able to plot all 84 banks and industrials into a three-dimensional joint space.[43]

Levine's first version of this plot is shown in Figure 3.6. Here, the banks are named and the industrial corporations are indicated by a number. Note that while Levine's data only contain bank-corporate ties, his procedure has effectively clustered banks and industrials by geographic *region*. To choose the example he does, the First National Bank of Chicago (right-center of the map) falls close to Northern Trust, Harris Trust, American National, and Continental Illinois Bank—all headquartered in Chicago. Major Chicago industrial firms—Standard Oil of Indiana (#15), Armour (#24), Inland Steel (#68), and International Harvester (#41)—are nearby. A similar Pittsburgh-based cluster appears just below the center of the plot (Mellon National Bank, Union National Bank, etc.). This *structural* fact—namely, that the American corporate-financial system is to a great extent geographically localized—is, of course, something which was *not* built into the mapping procedure, but is simply a feature of the network of directorship ties. Other research, however, confirms that this is an accurate portrayal of U.S. corporate organization.[44]

In the map shown in Figure 3.6, two dimensions are represented directly and a third is provided by separating the banks from the industrials. This plot, however, can be decomposed into two separate ones, as in Figures 3.7 and 3.8.

In Figures 3.7 and 3.8, we can see some of the difficulties that can arise when we try to reduce the dimensionality of interpoint distances in an N-dimensional array. In Figure 3.6 some industrials which are closely linked to banks in the original data—and which therefore ought to appear close to them in the plot—are quite far apart. Gulf Oil (#10), for instance, is *not* close to the Mellon National Bank, to which it is heavily tied.[45] Levine discovered, however, that these "discrepancies" formed a pattern: a bank's cluster does not have the same second- and third-dimensional location as the bank itself. By drawing lines from banks *through* industrials closely tied to them in the original data—as in Figures 3.7 and 3.8—Levine discovered that these lines converged at or close to a single point.[46] This means that the joint plot given in Figure 3.6 has a "center" and that

> the industrials and the banks lie on concentric shells of a sphere. . . . The separate clusters are organized around one center [so that] viewed spherically, there is a Morgan *sector*, a Chase *sector*, a Mellon *sector*. Radii indicate the strength of the link, angles indicate directions.[47]

The "discrepancies" in Figure 3.6, then, are really a function of what happens when drawing a three-dimensional object onto a two-dimensional surface. Geographers and cartographers, however, have devised some clever ways of *projecting* three-dimensional objects (the earth, for instance) onto two-dimensional surfaces. Levine uses one of these, called a *gnomonic projection*, to eliminate some of the

Figure 3.6

Bank-Industrial Joint Space

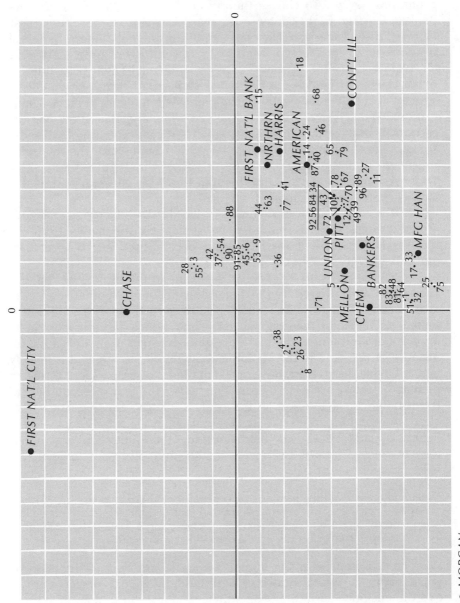

Source: Joel H. Levine, "The Sphere of Influence," p. 20. Reprinted by permission of the publisher.

Figure 3.7

Levine's Bank-Industrial Linkages: Dimensions 1 and 2

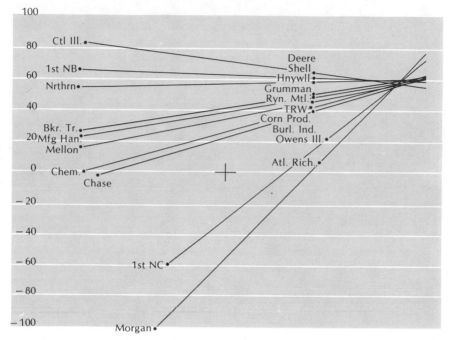

Source: Joel H. Levine, "The Sphere of Influence," p. 21. Reprinted by permission of the publisher.

discrepancies we saw in the map in Figure 3.6. This gnomonic mapping is given in Figure 3.9.

Gnomonic projections, Levine notes, have the advantage that all points on the same radius are mapped together. Great circle routes, that is, paths that follow a circle on the surface of a sphere whose plane passes through the center of the sphere, are converted into straight lines. Thus, the map given in Figure 3.9 is quite similar to what we would see in two dimensions if Levine's data were projected onto the ceiling of a planetarium: it tends to preserve the *relative* position of objects, but not their distance from the eye of the viewer.[48] The spaces formed by the intersecting coordinates in Figure 3.9, which appear "smaller" toward the center of the plot than at the periphery, cover the same measured area. The differences in their sizes are thus intended to convince the viewer that they are simply farther away.

While Levine's "Sphere of Influence" was published only three short years after Dooley's article, the difference in the methodological sophistication and subtlety of the two pieces is striking. Levine is able to

Figure 3.8

Levine's Bank-Industrial Linkages: Dimensions 1 and 3

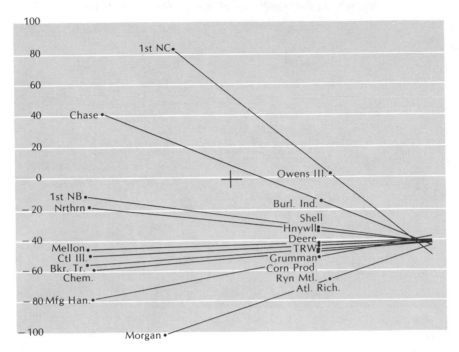

Source: Joel H. Levine, "The Sphere of Influence," p. 21. Reprinted by permission of the publisher.

solve the problem of the proportional effects of board sizes on interlocking by treating directorship ties as part of a single continuous network and then *generating* a metric on the basis of their *pattern* in this network as a whole. The distance between banks and corporations in both his two-dimensional and three-dimensional plots thus becomes a relative, rather than absolute, function of the raw number of directorship connections. Beyond this, Levine's gnomonic projection begins to suggest questions about *larger units*, that is, *sets of associated firms*, which might be worth examining.[49] What are the consequences for a firm, for instance, of falling within the "Chase sector" as opposed to the "Morgan" or "First National City Bank sector" (in Figure 3.9)? Are firms within the same bank-sector likely to compete? to draw upon the same capital pool? to have other kinds of nondirectorship ties in common? to share ownership?

Levine's article, then, clearly indicated some of the ways in which structural analysis might combine various scaling and mapping techniques in order to produce powerful portraits of the overall morphology of complex social structures. As we shall see presently, the style of work which it represented, as well as the specific techniques involved, quickly became the state of the art.

Figure 3.9

Gnomonic Map of Levine's Data

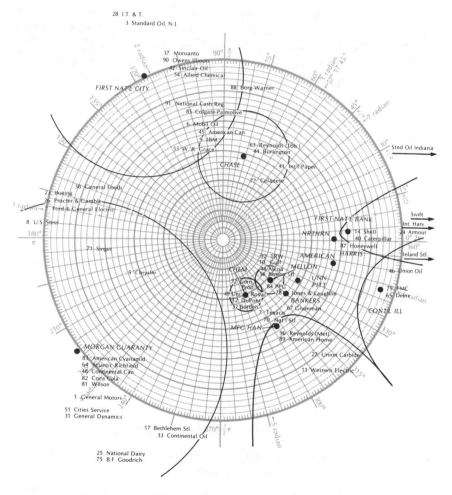

Source: Joel H. Levine, "The Sphere of Influence," p. 25. Reprinted by permission of the publisher.

Empirical vs. Theoretical Methods

For our purposes, scaling techniques of this kind have one important drawback: while plots may be held up to tests of consistency and parsimony, they cannot, in themselves, be falsified. Only models which combine descriptive or pattern-recognizing procedures, such as multidimensional scaling, with some definite set of propositions can be treated in this way. Thus, representations of data of the kind Levine has created must be *interpreted* or analyzed before they can be used as part of a causal research design.

There are two basic strategies which researchers may follow in doing this: *empirical* or *theoretical model building*. Since, in practice, these approaches are often quite different, and since a *strategy* of model building is not the same thing as a model itself (*all* models, in some sense, are "theoretical"), let us define these two terms formally.

Definition 6: Empirical and Theoretical Model Building

If

O_R = a discrete class of observed objects in the "real world"; such that $O_R = \{o_{r_1}, o_{r_2}, \ldots, o_{r_n}\}$

E_R = a discrete class of observed events in the "real world"; such that $E_R = \{e_{r_1}, e_{r_2}, \ldots, e_{r_n}\}$

O_D = a discrete class of objects as represented in a set of data; such that $O_D = \{o_{d_1}, o_{d_2}, \ldots, o_{d_n}\}$

E_D = a discrete class of events as represented in a set of data; such that $E_D = \{e_{d_1}, e_{d_2}, \ldots, e_{d_n}\}$

O_M = a discrete class of objects as represented in a theoretical model; such that $O_M = \{o_{m_1}, o_{m_2}, \ldots, o_{m_n}\}$

E_M = a discrete class of events as represented in a theoretical model; such that $E_M = \{e_{m_1}, e_{m_2}, \ldots, e_{m_n}\}$

ϕ = a homomorphic mapping function

Then, given (O_R, E_R) and (O_D, E_D), empirical model building requires that we find (O_M, E_M) and ϕ such that

$$\phi : (O_R, E_R) \rightarrow (O_D, E_D) \rightarrow (O_M, E_M)$$

And instead given (O_M, E_M) and (O_R, E_R), theoretical model building requires that we find (O_D, E_D) and ϕ such that

$$\phi : (O_R, E_R) \rightarrow (O_D, E_D) \rightarrow (O_M, E_M)$$

Here the mapping function, ϕ, to the left of the colon denotes the character of the operation represented by the arrows to its right. Note that the result of both strategies—as shown in the symbolic form of the mapping—is the same: both ultimately generate consistent "fits" between "observed reality," data, and models. An empirical research strategy, however, *starts* with data. Only after a researcher has gathered these data does he or she begin looking for models to "explain" them and a function which maps them into the model. A theoretical research strategy, by contrast, *begins* with a model which carefully specifies the conditions under which it can be falsified. An analyst then operationalizes his or her measures and begins seeking "data" which will fit that model and a function (ϕ) which will map them onto it. Thus, in practical terms, the strictures placed on "theoretical" model builders are greater than those on "empirical" modelers: the latter can, in effect, adjust *both* their models and their mapping functions in order to achieve consistency. But, particularly in the early phases of research into a given problem, it is sometimes impossible to *propose* theoretical models without first "getting the lay of the land." Under these circumstances, it is both legitimate and necessary to employ empirical model building strategies. While, in strict terms, scientists try to avoid empirical model building precisely because of this additional latitude for "adjusting" fits of models and functions to data, it is often useful to employ it in an "exploratory" way.

This is precisely what Levine and a number of other sophisticated users of multidimensional scaling techniques have been doing. Given the state of our basic knowledge at the time his study was written, empirical model building was needed in order to obtain a rudimentary outline of corporate structures. As we mentioned earlier, however, structural analysts also succeeded in developing a range of techniques which are more readily adapted to strictly theoretical model building.

Algebraic Modeling: The Basic Ideas

In Chapter 2, we introduced the idea that a kinship structure could be abstractly described by a set of simple and compound roles as seen from the point of view of ego. Specific kinship structures could then be modeled through concatenation of permutation matrices corresponding to prescribed marriage rules—and compared with actual or concrete kin systems.[50]

During the decade that followed the publication of White's seminal study, *An Anatomy of Kinship*, structuralists were able to generalize this approach to modeling complex and interrelated social structures. As we mentioned earlier, Boyd had extended and enhanced White's work by demonstrating how the concept of a *homomorphism* can be used to compare the mathematical properties of different structural models.[51]

This set the stage for a general exploration of the formal rules governing the behavior of the elementary units of a range of kinds of organized systems.

In 1971, Lorrain and White reported the development of an algebraic modeling technique which was unique in that it could simultaneously take into account the perspectives of all nodes within a system.[52] This was a significant advance over earlier methods because it allowed them to distill out regular features of the *global* organization of a set of elements and to represent them at higher levels of abstraction. By using their technique, a given researcher could fashion increasingly more parsimonious descriptions of structures at each of several successive layers within a system, but without "losing" essential detail about the positions of particular actors within it.

A Tale of Two Deans

To illustrate how this method works, let us examine a practical problem which many academic administrators in large universities appear to be facing these days. Two deans meet for the first time. Both have a large number of sub-deans working for them. Neither dean, of course, is exactly sure what all these people do. Both, however, are convinced that each sub-dean knows from whom he or she receives instructions and to whom he or she reports. Since all the sub-deans in this hypothetical university have fundamentally the same titles, each dean would like to know (a) whether or not the *positions* occupied by his or her staff form a coherent and interpretable structure and, if so, (b) which sub-deans are structurally equivalent to one another. In addition, both deans would like to compare the organization of their respective offices.

They resolve to conduct a survey. Data are gathered on each of two types of ties: *command ties (C)*, which show who sends instructions to whom, and *reports ties (R)*, which show who reports to whom. A consulting structuralist is called in to interpret these data according to the Lorrain-White method.

There are six sub-deans working for the first dean. Their relationships to each other and to the dean (node 1) are graphically depicted in Figure 3.10.

Solid lines in Figure 3.10 indicate the direction of flow of "commands" or instructions, and broken lines the direction of flow of "reports" or responses. Note that each command tie is reciprocated by a reports tie, and vice versa. These types of ties are shown in matrix form in Figure 3.11.

These two types of ties can then be used as what Lorrain and White refer to as "generators" for forming compound role relations (e.g., *CR,*

Figure 3.10

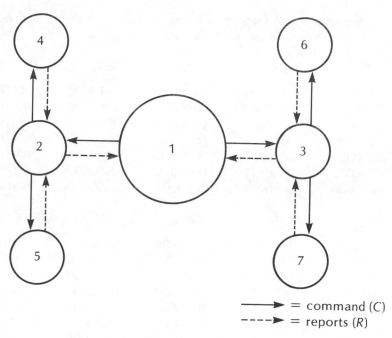

⟶ = command (C)
----▶ = reports (R)

Figure 3.11

Matrices Corresponding to C and R
Ties Shown in Figure 3.10

C	R
0110000	0000000
0001100	1000000
0000011	1000000
0000000	0100000
0000000	0100000
0000000	0010000
0000000	0010000

RCC, CRR) which reflect the composition of their corresponding patterns of ties with one another at various orders of interconnection or extension.[53] In this case, there are 14 elements in the semigroup algebra generated by these operations: *C, R, CC, RC, CR, RR, 0* (null), *RCC, RRC, CCR, CRR, RRCC, CRRC,* and *CCRR.* The matrices corresponding to these "words" are shown in Figure 3.12.

Figure 3.12

Elements in the Semigroup Formed from the C and R Relations
Shown in Figure 3.11

1:C	4:RC	7:000	10:CCR	13:CRRC
0110000	0000000	0000000	0110000	0000000
0001100	0110000	0000000	0000000	0110000
0000011	0110000	0000000	0000000	0110000
0000000	0001100	0000000	0000000	0000000
0000000	0001100	0000000	0000000	0000000
0000000	0000011	0000000	0000000	0000000
0000000	0000011	0000000	0000000	0000000

2:R	5:CR	8:RCC	11:CRR	14:CCRR
0000000	1000000	0000000	0000000	1000000
1000000	0100000	0001111	1000000	0000000
1000000	0010000	0001111	1000000	0000000
0100000	0000000	0000000	0000000	0000000
0100000	0000000	0000000	0000000	0000000
0010000	0000000	0000000	0000000	0000000
0010000	0000000	0000000	0000000	0000000

3:CC	5:RR	9:RRC	12:RRCC
0001111	0000000	0000000	0000000
0000000	0000000	0000000	0000000
0000000	0000000	0000000	0000000
0000000	1000000	0110000	0001111
0000000	1000000	0110000	0001111
0000000	1000000	0110000	0001111
0000000	1000000	0110000	0001111

The *composition table* showing the relationships between these elements within this algebra is given in Figure 3.13. The symbols in the table indicate the products formed by composing each of these 14 elements with one another. Where zeros appear in this table they indicate that the corresponding relation is null or empty. Note that relation 7 in the algebra is itself null so that all of its cross-products are null, as well.

Figure 3.13

Composition Table of the Semigroup Generated from the C and R Relations in Figure 3.11

	C	R	CC	RC	CR	RR	000	RCC	RRC	CCR	CRR	RRCC	CRRC	CCRR
C	CC	CR	000	C	CCR	CRR	000	CC	CRRC	000	CCRR	RCC	CCR	000
R	RC	RR	RCC	RRC	R	000	000	RRCC	000	CRRC	RR	000	RRC	CRR
CC	000	CCR	000	CC	000	CCRR	000	000	CCR	000	000	CC	000	000
RC	RCC	R	000	RC	CRRC	RR	000	RCC	RRC	000	CRR	RRCC	CRRC	000
CR	C	CRR	CC	CRRC	CR	000	000	RCC	000	CCR	CRR	000	CRRC	CCRR
RR	RRC	000	RRCC	000	RR	000	000	000	000	RRC	000	RCC	000	RR
000	000	000	000	000	000	000	000	000	000	000	000	000	000	000
RCC	000	CRRC	000	RCC	000	CRR	000	000	CRRC	000	000	RCC	000	000
RRC	RRCC	RR	000	RRC	000	000	000	RRCC	000	000	RR	000	RRC	000
CCR	CC	CCRR	000	CCR	000	000	000	CC	000	CCR	CCRR	000	CCR	CRR
CRR	CRRC	000	RCC	000	CRR	000	000	000	000	CRRC	000	RCC	000	000
RRCC	000	RRC	000	RRCC	000	RR	000	000	RRC	000	000	RRCC	000	000
CRRC	RCC	CRR	000	CRRC	000	000	000	RCC	000	000	CRR	000	CRRC	000
CCRR	CCR	000	CC	000	CCRR	000	000	000	000	CCR	000	000	CCR	CCRR

We can now answer one of the deans' questions: in the case of the first dean, at least, we can describe the structure of his or her office in a coherent way. In order to answer the second question they posed— whether or not some of the sub-deans were structurally equivalent to the others—however, it is necessary to introduce a hypothesis about the overall relationships between the relations reflected in the composition table of the algebra. The stronger this hypothesis—that is to say, the more demanding it is—the more likely it is to fail or be falsified. Since we can already determine one important thing about the network given in Figure 3.10, however—namely, that each type of tie is itself fully transitive—let us introduce the strong hypothesis that the result of *combining* them is *intransitive* and *commutative* (i.e., $CR = RC$).[54]

We can do this by *homomorphically reducing* the table, and the relationships associated with it, according to this principle. Two nodes will then be said to be structurally equivalent if and only if every relation between each of them and the other nodes in the reduced network is the same. The results of this reduction are given in Figure 3.14.

Figure 3.14

Results of Reduction of Types of Compound Relations Given in Figure 3.13

Set 1:	C	:	1, 8, 10
Set 2:	R	:	2, 9, 11
Set 3:	CC	:	3
Set 4:	RC	:	4, 5, 12, 13, 14
Set 5:	RR	:	6
Set 6:	000	:	7

We can see in Figure 3.14 that our original list of 14 elements has been reduced to 6. Elements 1 (C), 8 (RCC), and 10 (CCR) in the *original* algebra have now been mapped together into C. Similarly, elements 2 (R), 9 (RRC), and 11 (CRR) have been combined into R, and so forth. The new composition table, reflecting these reductions, is given in Figure 3.15.

Figure 3.15

Composition Table of Reduced Algebra

	C	R	CC	RC	RR	000
C	CC	RC	000	C	R	000
R	RC	RR	C	R	000	000
CC	000	C	000	CC	RC	000
RC	C	R	CC	RC	RR	000
RR	R	000	RC	RR	000	000
000	000	000	000	000	000	000

By mapping together all *nodes* that are structurally equivalent, we can order them into three sets, as follows:

Figure 3.16

Set 1: Dean: 1
Set 2: Sub-deans: 2, 3
Set 3: Sub-deans: 4, 5, 6, 7

Here we can see that the dean is structurally equivalent to him/herself alone. The sub-deans, however, have been divided into two sets of structurally equivalent nodes. By applying reduced relations to these sets of structurally equivalent deans and sub-deans, we can now describe the pattern of relations between them. Figure 3.17 shows the matrices corresponding to each of the types of relations in the *reduced* composition table (Figure 3.15).

Figure 3.17

1:*C*	4:*RC*
010	100
001	010
000	001

2:*R*	5:*RR*
000	000
100	000
010	100

3:*CC*	6:*000*
001	000
000	000
000	000

The mapping in Figure 3.17 is graphically represented in Figure 3.18. Note that in Figure 3.18 each set of nodes has an *identity relation* defined onto it. This indicates that the reduction was successful and that no anomalous ties occurred in the process.[55] If our reduction had not been perfect, one or more of these ties would not have been generated in the reduced algebra.

Figure 3.18

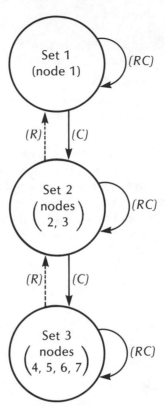

As Figure 3.18 clearly shows, our first dean had nothing to worry about. His or her office is structured into three tiers, with one set of sub-deans and one of sub-sub-deans. On the basis of this analysis, he or she could go back and redraw the network given in Figure 3.10 in a less confusing way, as in Figure 3.19.

The second dean in our hypothetical tale, however, is less enthusiastic: he or she has 14 people working for him or her. Is it possible that his or her office will be as neatly arranged as the other dean's? To test this proposition, we subject the raw matrices of C and R relations gathered from the second dean's staff to the same procedures as we did for the first. These data are presented in Figure 3.20.

By inspection, it is clear that the patterns of binary entries in the matrices in Figure 3.20 are similar to those we generated for the first case. The semigroup algebra in this instance, however, contains 30 elements and the composition table would be too large to print here. Pursuing exactly the same reduction strategy, however, yields the sets of structurally equivalent nodes shown in Figure 3.21.

Figure 3.19

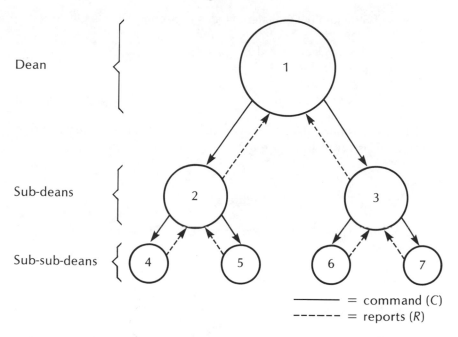

Dean

Sub-deans

Sub-sub-deans

―――――― = command (C)
------ = reports (R)

Figure 3.20

C and R Relations for Second Dean's Office

C	R
011000000000000	000000000000000
000110000000000	100000000000000
000001100000000	100000000000000
000000011000000	010000000000000
000000000110000	010000000000000
000000000001100	001000000000000
000000000000011	001000000000000
000000000000000	000100000000000
000000000000000	000100000000000
000000000000000	000010000000000
000000000000000	000010000000000
000000000000000	000001000000000
000000000000000	000001000000000
000000000000000	000000100000000
000000000000000	000000100000000

Figure 3.21

Sets of Structurally Equivalent Nodes in Second Dean's Office

Set 1: Dean: 1
Set 2: Sub-deans: 2, 3
Set 3: Sub-deans: 4, 5, 6, 7
Set 4: Sub-deans: 8, 9, 10, 11, 12, 13, 14, 15

The composition table, corresponding to the reduced algebra in this second case, is given in Figure 3.22.

Figure 3.22

Composition Table for Reduced Algebra in Second Case

	C	R	CC	RC	RR	CCC	RRR	0000
C	CC	RC	CCC	C	R	0000	RR	0000
R	RC	RR	C	R	RRR	CC	0000	0000
CC	CCC	C	0000	CC	RC	0000	R	0000
RC	C	R	CC	RC	RR	CCC	RRR	0000
RR	R	RRR	RC	RR	0000	C	0000	0000
CCC	0000	CC	0000	CCC	C	0000	RC	0000
RRR	RR	0000	R	RRR	0000	RC	0000	0000
0000	0000	0000	0000	0000	0000	0000	0000	0000

The matrices corresponding to each of the elements in this table are shown in Figure 3.23. Note that in Figure 3.23 the CR relationship defines identity relations onto each of the sets of structurally equivalent nodes. Thus, we strongly suspect that our reduction here was successful as well. The reduced network, representing the second dean's office, is given in Figure 3.24.

The second dean, then, has an office which is structured in the same way as the first's, except for the fact that it contains one additional tier. By algebraically reducing the C and R relations in the same fashion in both cases, the *structural similarities* among the positions occupied by personnel in both offices become clear. Had these two offices been structured in different ways, moreover, the procedure we followed here would have indicated the differences between them.

Figure 3.23

1:*C*	5:*RR*
0100	0000
0010	0000
0001	1000
0000	0100

2:*R*	6:*CCC*
0000	0001
1000	0000
0100	0000
0010	0000

3:*CC*	7:*RRR*
0010	0000
0001	0000
0000	0000
0000	1000

4:*RC*	8:*0000*
1000	0000
0100	0000
0010	0000
0001	0000

Implications for the Study of Economic Structure

As this "tale of two deans" indicates, the principal strength of the Lorrain-White method and later algebraic techniques lies in their ability to deal, simultaneously, with general roles or role structures and with the particular locational properties of nodes ("positions"). The examples we chose to illustrate this feature of algebraic modeling were, of course, quite straightforward. In fact, one could detect an orderly, hierarchical pattern in these data by inspection. More complicated "deep" structures, however, can often be uncovered using the same methods as we used in these examples.

If instead of "deans" or "offices" our nodes had been *corporations* and our relations directorship interlocks, the analysis of these data could have been carried out in a similar, but not identical, fashion (due to the fact that bonded-ties are, by definition, bidirectional).[56] The meaning of the notion of a "role" or "position," however, would have been less

Figure 3.24

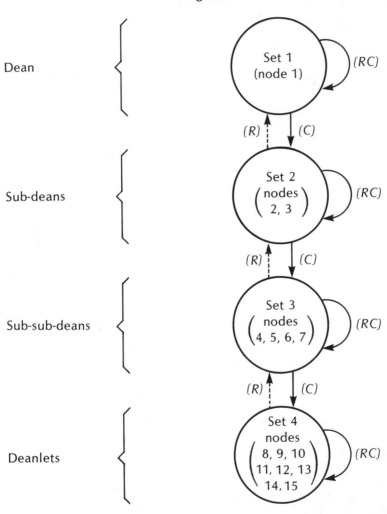

intuitively obvious: normally, we do not think of corporations as playing roles or occupying positions in the same way people do. The principle, however, is the same: the structural role played by a corporation and its position *within a structure* are theoretically analogous to those which standard sociologists discuss in social groups.[57]

For instance, let us assume that the dean in our first example was one of the major automobile companies and the sub-deans were *independent* parts manufacturers. If our measure of interconnection were directorship interlocks, and the structural pattern were the same as that shown in Figure 3.19, we would be justified in viewing corporations within each of the three tiers as playing discriminately different "roles" within the system of interlocking directorates. Similarly, *structurally equivalent* corporations could be determined on the basis of the relations between each node and the others within the larger net.

A *substantive* interpretation of the hierarchical pattern in these data, of course, would depend on a number of things—only one of which could be this relatively simple structural map itself. To note that a pattern of corporate interlocking is hierarchical, or that certain nodes within it are structurally equivalent, is not the same as explaining *why* it has assumed this particular form or what impact this morphology might have on the *behavior* of given corporations. Firms, after all, are not people, but organizations, and we need some sort of more specific theoretical framework to explain how they might behave under specified structural conditions.

INTERCORPORATE TIES AS ECONOMIC STRUCTURES

While scaling techniques and algebraic modeling made it possible for structuralists to begin creating explanatory schema of this kind, then, by the mid-1970s it was clear that these methods, in themselves, would not provide ready answers. Too many aspects of the problem of abstracting or modeling corporate structure were poorly defined at the outset, and the particular types of intercorporate ties which structuralists had focused on, namely, interlocking directorships, were much more easily mapped than interpreted.

As we mentioned earlier, structural analysts have historically pursued one of two lines of attack in challenging the conventional view of economic systems as abstract arenas for exchanges among independent actors. The first had largely grown out of exchange theory and studies of social organization, and the second out of a particular Marxian interpretation of the role of financial and fiduciary institutions in modern corporate systems. In the earliest stages of structuralist research in the area, the borders between these two had been blurred. Levine's study, for instance, while formally simply a mapping exercise, was implicitly rooted in both: his assumption that *multiplexity* implies "closeness" suggests that some form of communication or integration is at work (exchange theory/social organization framework), and his discrimination of nodes into banks and industrials reflects an attempt to examine Hilferding-like models.[58] Later studies, however, tended to diverge and follow one or the other of these analytic streams.

Thus, as time went on, there was an increasing tendency for structuralists pursuing these problems to borrow techniques from one another, but little else. Two major research projects exemplify how these divergences came about.

The Toronto and Stony Brook Projects

In 1973-74, structural analysts at two North American universities began assembling large-scale corporate data bases. Initially, both groups were primarily interested in directorship interlocking. Even at the outset, however, the "Toronto Project" had intended to associate whatever morphological features of directorship data it might discover with behavioral "outcomes" (e.g., measures of market activity, debts, profits).[59] The "Stony Brook group," by contrast, was chiefly concerned with the coordination or control of economic structures.[60]

These divergent interests are reflected in the kinds of information each group gathered. In addition to directorship interlocks, the Toronto research team systematically collected data on a wide range of forms of interaction between corporations (e.g., intercorporate ownership, borrowings, common participation in markets).[61] The Stony Brook group largely confined itself to directorship patterns, since (a) data on stockholding is not easily available in the United States, and (b) given its overall research design, extensive behavioral measures were not necessary.[62]

Both groups quickly discovered an important fact which most earlier researchers had not, probably because their "top down" corporate samples were too small: patterns of directorship interlocking are typically not only dense and intricate, but usually do not contain easily "detected" cliques and clusters.[63] Most earlier researchers, because they had collected and analyzed egocentric networks, had simply assumed that larger networks of directorship ties would be structured in this way. It quickly became clear, however, that when bigger samples are drawn, apparently distinct egocentric networks may merge into a few completely connected components within larger graphs.[64]

For instance, the Toronto group discovered that 4,101 of the 5,306[65] firms associated with the 361 largest public, private, and Crown corporations operating in Canada in 1972 shared at least one ownership tie $\geq$ 10 percent with one another at some remove; that is, where t = a tie $\geq$ 10 percent and $t \equiv t^N$ for all such t. Since the average number of ties sent out by a given firm in this larger set was approximately equal to 2, these results closely coincided with those expected at random.[66] Similarly, when director/officership ties (which mapped over these ownership ties) were treated as binary, they also formed substantially random patterns—as measured by a slightly modified version of Rapoport and Horvath's classic "tracing" procedure. Since the information value of a random measure is 0, even if these two measures had been combined, without some modification the resulting pattern would have been largely random. More important, since *parts* of a random net may be *locally* nonrandom, attempting to generalize about "groups" or clusters of interconnected firms *within* such a net would have been extremely hazardous.

While members of each of these research teams dealt with this problem of global randomness or intractability somewhat differently, on balance the Toronto research team reacted by intensifying its search for *specific units* within this larger structure which could be associated with given behavioral outcomes. By contrast, the Stony Brook group tended to look to problems which were susceptible to global measurements taken on a network as a whole, for example, centrality. In effect, the Toronto group—which was strongly influenced by the exchange theory/social organization tradition—tried to devise specific theoretical models to "detect" discriminate structures *within* this random net. The Stony Brook group—which was strongly influenced by the financial-control model—simply avoided the issue by focusing on problems where the global *patterns* of ties *among specific sets* of corporations were not directly relevant. Interestingly, while the directions pursued by these two groups were quite discernibly different, the results they obtained were not only compatible, but complementary.

The principal investigators in the Toronto group—Stephen Berkowitz, Yehuda Kotowitz, and Leonard Waverman—began by exploring a number of ways of sharpening the standard definitions of *firms*, *enterprises*, and *establishments*. Apart from a few specialists in "industrial organization," most economists use the first two of these terms interchangeably. A firm, they assert, is simply a legal persona that corresponds to a functioning business or enterprise.[67] Establishments, in this framework, are the actual physical entities through which a given enterprise carries on its affairs. These relationships are graphically represented in Figure 3.25.

In Figure 3.25 the double arrow indicates that given a particular function (interpretation), enterprise α can be mapped into firm α, and vice versa. Establishments A and B, however, are contained within enterprise α; they are simply its "operational arms."[68]

In practice, this conventional typology breaks down rather badly. Note, for instance, that it is utterly incapable of dealing with any number

Figure 3.25

Conventionally Understood Relationships between Establishments, Firms and Enterprises

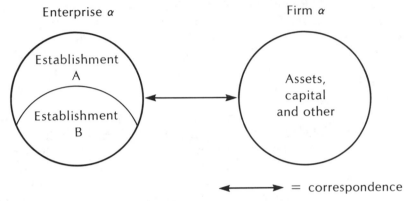

Enterprise α

Firm α

of common "joint marketing" or "joint production" arrangements in which more than one enterprise operates or more than one firm owns part of a single establishment. True "joint ventures," that is, firms owned by more than one other firm, can be accommodated, but only in a very cumbersome way,[69] as illustrated in Figure 3.26.

Figure 3.26

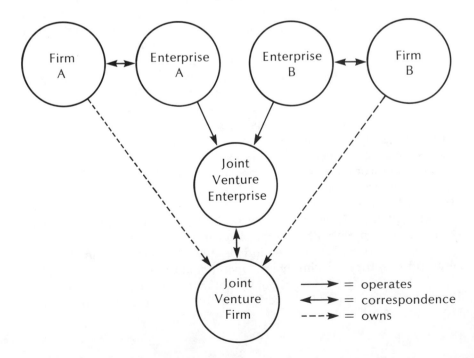

3: Corporations and Privilege

More routine difficulties occur whenever someone tries to employ this schema in dealing with subsidiaries. Since they are legally incorporated, subsidiaries are firms. In conventional terms, they would have to be treated as enterprises or as some kind of "special" case. This creates problems for government statisticians, who use "enterprises" as units of analysis in calculating "market share" or capital leverage, since subsidiaries do not operate independently. Thus, many government economists have begun reckoning the assets, shipments, and so forth, of *majority*-owned subsidiaries as part of those of their "parent" firms.[70]

Berkowitz et al. formalized this practice and generalized it to deal with a range of other legal and informal patterns of control among nominally "independent" corporations. Following the lead of other structuralists, they created a *three-tiered* model in which each tier corresponded to a layer or level of organization. This is shown in Table 3.1.

Table 3.1

Relationships between Levels of Modern Corporate Systems

Level	Unit	Basis of Definition	Function
1	Establishment	Physical or geographic contiguousness of capital assets	Physical locus for production, provision of services, employment, etc.
2	Firm	Legal requirements for a corporate persona through which business can be organized	Holding of capital or other assets; party to contracts; incur debts; acquire and dispose of assets, etc.
3	Enterprise	Some measure of power or, in the limiting case, control exercised between and among firms	Allocate capital among alternative uses; garner and assign credit; coordinate management or production

Source: S. D. Berkowitz, P. J. Carrington, Yehuda Kotowitz, and Leonard Waverman, "The Determination of Enterprise Groupings through Combined Ownership and Directorship Ties," p. 395. Reprinted by permission of the publisher.

Note that in Table 3.1 each layer consists of units which may be mapped into those at the level just below it in the typology. Establishments, for instance, can be ordered together on the basis of their legal owners, firms. Firms, in turn, can be grouped together into enterprises. In the limiting case—a "mom and pop" grocery store, for instance—all three levels would coincide, that is, a given unit can be a single establishment-firm-enterprise. Under other circumstances, each level of organization may be distinguished from the others.

By typologizing corporate structures in this way, Berkowitz et al. made it possible for researchers to use graph theory to tease strongly interrelated clusters of firms out of the maze of interconnections in which they are embedded. Their own contribution to this task was a new way of calculating what mathematicians refer to as *transitive closure*.

Under normal circumstances, transitive closure is calculated by simply adding together (under Boolean arithmetic) the successive powers of a matrix representing the ties between nodes in an acyclic directed graph or *digraph*.[71] As we noted in Chapter 2, raising a matrix to its powers corresponds to *reducing* the length of the distances between nodes. Thus, given a matrix $G = [g_{ij}]$, $g^n_{ij} \# =^1 1$ iff there is path of length n from vertices (nodes) v_i to v_j and

$$\left(\sum_{k=1}^{n} g^k_{ij} \right) \# = 1,$$

iff there is at least one path of length $\leqslant n$ from vertices v_i to v_j.[72]

Where there are n vertices, this summation is complete at the $(n-1)^{\text{th}}$ power of a matrix. Given an acyclic digraph, any path of $q > 1$ contains one of length $q - 1$. Thus, it is sufficient to sum to the qth power where

$$\left(\sum_{k=1}^{q+1} G^k \right) \# = \left(\sum_{k=1}^{q} G^k \right) \#$$

since for all m, $m > q$

$$\left(\sum_{k=1}^{m} G^k \right) \# = \left(\sum_{k=1}^{q} G^k \right) \#$$

if we exclude cases were G^{q+1} is nonzero but contains paths duplicated in $G^1 \ldots G^q$.[73]

This normal notion of transitive closure is inadequate in the case of corporate networks, because if one is interested in ordering firms together into enterprises, that is, groups of firms operating under common control, one must be able to deal with the case where several firms combine in order to control others, for example, "joint ventures." In order to solve this problem, Berkowitz et al. created a *value matrix*, *M*, which represented one or another of their measures of control between firms. By raising this matrix $M = [m_{ij}]$ to successive powers, "summing the powers together in a stepwise fashion and binarizing (changing all values greater than the control criterion to 1 and all others to 0) each successive power of the matrix before proceeding to the next multiplication and addition," it was possible to model the fundamentally *binary* nature of corporate control: a given firm or set of firms either controls another or it does not. At each stage in the process, "cycles of length $n - 1$ were removed by changing all nonzero diagonal elements to 0."[74]

In this fashion, Berkowitz et al. were able to locate enterprise "parents," and "heads of joint ventures." All other firms, v_i, were designated as members of enterprises and joint ventures and were uniquely assigned to the enterprise parent (or joint venture head), v_j, having the greatest summed holding in v_i (i.e., v_j where

$$m_{ji} \in \sum_{k=1}^{n} = 1 \ M^k \ \#\#$$

was greatest)."[75]

With this tool in hand, Berkowitz and his colleagues were able to explore the implications of a number of different measures of "control" for (a) the boundaries among multifirm enterprises, and (b) the boundaries between given multifirm enterprises and less closely tied sets of firms—in effect, their "environment." The results of this research are reported in two studies, *Enterprise Structure and Corporate Concentration* (1976), and "The Determination of Enterprise Groupings through Combined Ownership and Directorship Ties" (1979).[76] These are too lengthy to be discussed here in detail. However, Berkowitz et al.'s work raises a number of points which bear directly on the larger problem of measuring and interpreting corporate structure.

First, conventional definitions of interfirm control are clearly inadequate. The two *control criteria* normally used by economists, namely, ownership of 10 percent or more (Larner) or ownership of 20 percent or more of voting stock (Berle and Means)—which are based on the principles of "leveraged ownership" first proposed by Hilferding— lead to trivial or inconsistent mappings. As we noted earlier, the application of the ≥ 10 percent criterion to Canadian corporate data for 1972 yielded random clusters. The effects of the "20 percent or more" criterion (a) do not substantially differ from those produced by the "15 percent or more" measure and (b) are difficult to justify, in themselves, since the corresponding enterprise arrays can be disturbed in a relatively easy manner by making or breaking ties.[77] Obviously, the use of overall ownership criteria as an exclusive measure of control between firms is likely to produce poorly differentiated clusters.

Second, when similar tests of consistency, robustness, and so forth, are performed on director/officership matrices, and multiplexity is used as a measure of similarity or adjacency of nodes, the results are also inconclusive.[78] While, as in the case of pure ownership criteria, nontrivial clusterings may be generated by using some highly restrictive measure of interconnection (e.g., at least one ownership tie *and* three or more directorship interlocks), it is difficult to interpret these patterns, in themselves.

Third, the use of combined director/officership and ownership criteria makes it possible to create nontrivial and robust clusters of enterprise arrays. The results produced by using executive board interlocks as part of a definition of control—as a number of economists suggest—are inconsistent and strongly imply that executive board membership ties serve some other purpose.[79]

Finally, while in the Canadian case the use of combined measures of control between and among firms produces marked results in terms of the sheer size of enterprise groupings, it does not fundamentally alter the *behavioral* outcome in which Berkowitz et al. were interested, namely, top-4 and top-8 concentration ratios.[80] Economists use these ratios to index the degree to which a given market is dominated by the largest firms operating within it. With the exception of 9 of 160 industry categories, Berkowitz et al. discover that market concentration is unchanged when firms are ordered together into larger enterprises on the basis of *minority-control criteria*. This suggests that—once again in the Canadian case—*horizontal integration* (ties among firms operating in the same product market) is either brought about through *majority* stock ownership or is simply uncommon.

The Stony Brook group, headed by Michael Schwartz, focused on broader political economic issues similar to those dealt with by Aaronovitch and Barratt Brown. Given a modern industrial structure, it asked, is it possible to determine precisely where power is located within it and which people or institutions wield this power? While these questions were similar to those posed in earlier studies, the methods the Stony Brook group employed in answering them were quite different.

Social scientists doing power-structure research had traditionally been reluctant to formulate precise questions or to subject their "hypotheses" to close empirical tests. The literature, as a result, is filled with half-assertions backed up by selectively chosen "facts." The hallmark of the Stony Brook group, by contrast, was its willingness to address "big issues" of the same kind, but with clearly formulated formal models and carefully gathered data sets.

An article published by the group in 1976—"Problems of Proof in Elite Research"—unambiguously stakes out this position. It criticizes earlier power-structure research, including some conducted by its own members, for failing to devise *critical experiments* that would distinguish between alternative models of power holding.[81] It underscores the importance of falsifiability in research of this kind and suggests that the unresolved debate between pluralist and antipluralist writers might be well served by more careful attention to this aspect of modeling.

Moreover, in this paper and in later work, members of the group stress the subtle interdependence of models and the observations used to "test" them. In summarizing previous research on elite structures, Beth Mintz, Peter Freitag, Carol Hendricks, and Michael Schwartz conclude that:

> The democracy/dictatorship debate has a peculiarly elastic quality. Given any finite set of facts about American politics, it is possible to construct a pluralist theory which fits them. The same set of facts can also be fitted to a Marxist or power elite theory. This makes comparison of the theories difficult, but not impossible. Carefully defined research can be designed critically to test the most plausible version of pluralism against the most plausible anti-pluralist model. . . . Researchers must pay close attention to theory construction, to the creation and expression of complete and sophisticated theories which admit to detailed analysis and empirical test.[82]

This combination of a broad interest in political-economic issues and an ability to focus in on specific models and data sets shows through clearly in two studies of American corporate structure conducted by the Stony Brook group: "The Nature and Extent of Bank Centrality in Corporate Networks" and "The Role of Financial Institutions in Interlock Networks."

Both papers use corporate interlocks as an index of integration among firms, and both largely rely on centrality measures as a means of distinguishing among the roles of particular firms or groups within a larger structure.[83] In the earlier study, the authors try to uncover natural clusters of firms which coincide with some discriminant property of nodes (e.g., their location, industrial type). If the pattern formed by all directorship interlocks is examined, however, there is little evidence of groupings of this kind. This result, of course, parallels those obtained by the Toronto group. If only those interlocks attributable to directors who are also officers of a corporation (director/officership ties) are considered, however, clear *regionally based* clusterings emerge which are extensively but loosely connected into a single national network.[84]

Since the later study expands upon and refines these findings, let us look at it in some detail. The key measure employed by the Stony Brook group in its research is a variation on the centrality indices we discussed in Chapter 1. The members of the group reasoned that the location of a firm within a corporate network reflects its position within a structure for the mobilization of resources and is indicative of its specific role within a market or markets.[85] Firms with high centrality, they argued, occupy "pivotal positions" and therefore tend to dominate other market participants in their immediate areas.

Centrality measures, as we saw earlier, are primarily designed to reflect positional aspects of nodes within a system of relations. The earliest centrality measures, which were based on the degree of nodes, only indirectly took into account the global pattern or dispersion of scores among the points of a network.[86] Later measures attempted to embody global properties of the network in their indices of point centrality in a more direct fashion.

The members of the Stony Brook group focused on one such index, developed by Phillip Bonacich, which had been designed to deal with overlapping memberships between and among organizations.[87] In Bonacich's measure, it is first necessary to eliminate the effects of the sheer size of a network on the pattern of overlap between nodes. Then centrality itself can be calculated independently of these effects. Bonacich shows that this first step may be accomplished by multiplying a matrix reflecting the overlap between groups by a series of positive numbers, which has the effect of artificially introducing a "standard" group size for all groups.[88] The resulting matrix of standardized overlap scores can then be directly manipulated to yield a vector of centrality scores where the overlap of one group with others is weighted by the centrality of the groups to which it is connected.[89]

Bonacich's technique therefore made it possible to examine not only the specific graphic interconnection between nodes, but the *potential* for interconnection implicit in the opportunities for overlapping of personnel at those points. It also yielded a standardized centrality index which takes into account all centrality scores simultaneously.

The members of the Stony Brook group then modified Bonacich's centrality index to suit their particular purposes. In modified form this centrality index is sensitive to three parameters: (a) the number of nodes (firms) with which a given node interconnects (interlocks), (b) the intensity of those ties, and (c) the centrality of the firms to which it is connected. Formally:

$$C_i \sim \sum_{\substack{j=1 \\ j \neq i}}^{N} r_{ij} * C_j$$

where: N = number of nodes

C_i = the centrality of the node under consideration

r_{ij} = the intensity of the connection between nodes i and j and $r_{ii} = 0$

C_j = the centrality of node j.

The key problem here is, of course, to define r_{ij} or the "intensity" of the connection between corporations i and j. If the centrality of the node under consideration is proportional to the sum of the products of the intensity of connections to receiving nodes and their respective centrality scores, then this intensity measure is critical. The Stony Brook group experimented with several of these.[90] For various reasons which are not important here they concluded that r_{ij} (a) should reflect the multiplexity or "multistrandedness" of the connections between nodes i and j, and (b) should be standardized to reflect the fact that as corporate boards increase in size, the influence of each director diminishes, that is, the impact of an interconnection on a given board is reduced. Thus, they then assumed that:

$$r_{ij} = \frac{b_{ij}}{\sqrt{d_j}}$$

where b_{ij} = the number of interlocks between corporation i and j

and d_j = the number of directors on the board of corporation j.

Given their application of Bonacich's centrality measure, then, it is important to take into account the context of corporate decision making, as well as the crude multiplexity of a node's connections. While the technique of dividing this by a square root measure in order to standardize it is reasonable and, to some extent, justified empirically, it is nonetheless arbitrary but mathematically correct, and other measures might be devised which would be as correct or more appropriate.[91]

In any event, it is clear that they proceed by assuming that (a) all ties are, initially, equally important, and (b) the impact of an in-coming tie on a given node is a function of the square root of its board's size. Thus, while their measure is somewhat sensitive to the "proportional effects" we discussed earlier, only grossly discrepant scores would have much of an impact on centrality as they measure it here.[92]

Mintz and Schwartz apply this measure to the patterns of interlocking directorship among samples of 1,131 and 1,111 companies drawn in each of two years, 1962 and 1966, respectively.[93] Their results demonstrate a clear tendency for financial firms to dominate the list of most central firms. Thus, in 1962 "nine of the ten most central firms and 13 of the highest 20 were financial concerns."[94] The regional pattern shown in the previous study tends to persist as well: "A majority [of these firms] were the established Northeastern financiers which dominated the banking industry since the turn of the century."[95] Industrial firms that fell into the "most central" category "varied in size, location, and industry type" but were generally less central than the financial institutions within this set.[96] Similar results were obtained for the 1966 data.

On the basis of a more detailed analysis of these findings, Mintz and Schwartz are then able to distinguish two structurally discriminant roles played by highly central firms—*hubs* and *bridges*.

> A hub is a corporation in the center of a group of interlocked firms, while a bridge is a company which links two or more hubs. Since corporate centrality is based on the centrality of all firms to which a company ties, bridges may be given high scores . . . simply due to connections to a few strategically located companies. Hubs on the other hand, obtain high centrality as a result of the cumulative weight of numerous ties to less central firms.[97]

This leads Mintz and Schwartz to discriminate between two forms of centrality—*hub centrality* and *bridge centrality*. The substantive implications of the two, they argue, are clear: hub centrality reflects the degree to which a given firm plays an active coordinative role within the structure and bridge centrality a passive or "go-between" role. On the basis of their data, Mintz and Schwartz conclude that American financial institutions predominantly play the first of these.[98]

While studies of the kind undertaken by the Toronto and the Stony Brook groups were far from conclusive, then, by the late 1970s structural analysis had taken some strong steps toward the classes of models, techniques, measures, and datasets necessary to test sophisticated propositions about the organization of economic systems. A clear distinction, however, could be drawn between researchers whose work reflected a primary interest in describing and analyzing corporate structure, per se, and those who saw these studies as simply one avenue that might be pursued in dealing with a range of more general issues related to power holding in advanced capitalist societies.

Both groups had succeeded in modest ways in doing what they set out to do. Structural analysts working within the exchange theory/social organization tradition had been able to associate gross patterns of director/officership ties with classic or conventional measures of corporate conduct. The Toronto group, for instance, had discovered (in the Canadian case) that while the size (number of firms) of the effective units used in measuring corporate concentration could be altered dramatically through the use of techniques for detecting groups of allied firms (enterprises), top-4 and top-8 market concentration ratios were essentially unaffected by regrouping them in this way. While there are probably, if not unique, at least unusual reasons why the Toronto group obtained the results it did—including the high "initial" levels of concentration in most Canadian industries—this *negative* finding is extremely important in that it suggests that *horizontal* integration of firms (as reflected in the horizontal measures of corporate concentration used by economists) is relatively uncommon. This behavioral result, moreover, is consistent with those obtained by another structuralist,

Ronald Burt, who showed that interlocking directorates in the United States are usually forged between firms in those *vertically* related industrial sectors which act to constrain each other's profit levels. Taken together, these findings seriously call into question the commonplace assumption that the principal purpose, and hence principal effect, of directorship ties is to limit competition or to create anticompetitive combinations *within* industries and product markets.[99]

Structural analysts who were previously interested in the relative power of financial and industrial corporations—and, by implication, in the relative ascendency of financiers and industrialists within advanced capitalist societies—had been able to establish the clear preponderance of financial and fiduciary institutions among the firms most central to patterns of interlocking directorates. The Stony Brook group, for instance, had shown that 9 of the 10 most central U.S. firms in 1962, and 13 of the leading 20, were financial intermediaries of precisely the kind Hilferding had predicted would dominate the mixed financial-industrial structure coming into being at the time he wrote.[100] Beyond this, members of the Stony Brook group had been able to distinguish between two discriminant roles played by highly central firms—hubs and bridges—which they showed were most often played by financial and industrial corporations, respectively. In addition, they found strong evidence of regionally based financial and industrial groupings in the U.S., which was consistent with the results obtained by other structuralists—such as John Sonquist and Thomas Koenig, and Levine—who had employed very different techniques in modeling their data.[101]

Neither of these branches of structural analysis, however, had succeeded in developing methods that would allow researchers to predict behavioral outcomes directly from measures of the morphology of corporate structures. Thus, while their findings clearly met conventional standards of proof, namely, statistical inference, structural analysts working on these problems were far from satisfied with the state of the art. Their goal had been to model the constraints under which elements at different levels of these structures operated. By the end of the decade, of all those working in the area, only Burt had tried to tackle this issue directly. While his research strongly suggests directions that can be followed in developing general models of constraint, his approach is far too restricted in its present form to be used in this way.[102]

CORPORATE SYSTEMS AND PRIVILEGE

During the 1970s, structural analysis had made useful theoretical and technical contributions to the description of complex patterns of corporate organization. Neither its "social organization" nor its Marxian branch, however, had succeeded in tying its techniques together into a clear structural framework for interpreting the *behavior* of the elementary parts of corporate or social systems. Thus, for instance, while the Toronto group could demonstrate that its notion of an "enterprise" is more consistent and parsimonious than conventional ones, it could not show what direct effects these units have on the market behavior of firms.[103] Similarly, while the Stony Brook project had produced a precise measure of corporate centrality, it had created only a very general model to explain the impact of central location either on power within economic systems or on the distribution of rewards and benefits in capitalist societies.

Clearly what was needed was some formal means of translating structural morphology into systems of constraints whose effects on behavior could be predicted. Game theory deals with precisely this idea: "behavior under constraint." But apart from one excellent article by Boorman, structuralists have tended to skirt the extremely complex technical problems involved in translating between these domains.[104]

Some structural analysts, however, have created less formal models which begin to address issues of this kind. The most developed of these focus on the broad relationship between economic structure and class or *privilege*.

In the mid-1970s, a number of researchers published studies which maintained that modern capitalist *upper classes* did not simply "emerge," but came about as a result of a long-term historical process in which sets of loosely interconnected families were gradually formed into an overarching and tightly interconnected structure. Each stage in this process, they argued, was strongly associated with a major change in the global organization of the production and distribution of commodities.[105]

Initially, merchant capitalists forged alliances through marriage ties and partnership arrangements. As the scale of their activities grew, however, they created incorporated bodies to act as capital pools for joint investments and as institutional foci for managing them.[106] The consolidation of industrial and, especially, financial firms into a single interconnected network of the kind we described earlier then facilitated the integration of these separate groups into a fully formed social class.[107]

Throughout this process, dynamic interaction went on between the kinship structure of privileged families, on the one hand, and increasingly sophisticated institutional arrangements for the mobilization of capital, on the other. During the earliest phases of this process, the consolidation of what Berkowitz refers to as *family compacts* established capital bases which were then institutionalized through the creation of family-dominated companies. These firms, in turn, facilitated the further consolidation of the families associated with them into larger units (*elites*), which, as a result, became institutionally tied to the central dynamics of the developing industrial system. This type of elite family organization also encouraged rivalries between coalitions of firms operating in financial markets. The creation of institutional mechanisms to settle, or at least manage, these conflicts eventually produced a single, loosely interconnected kinship group coterminous with an upper class.[108]

A number of researchers who study the structural transformation of privileged or propertied groups refer to this process as *structuration*.[109] At least two analysts—Berkowitz and Daniel Bertaux—try to describe it formally using elementary graph theory.[110] Both of these writers, moreover, draw an intentional parallel between this transformation of disjoint groups of merchant capitalists into a dominant upper class and the kinds of changes which occur when chemical elements, such as hydrogen and oxygen, combine or take on different physical states, namely, gases, liquids, solids. Bertaux, for instance, argues that in France

> the structure of capital . . . is no longer a structure of isolated molecules (family enterprises) but a structure of large groups, which are not dominated by one or two families but in which many families possess a small part.[111]

This change in the ability of various families to dominate or control the process of capital mobilization, Bertaux contends, is not simply a function of the scale of the activities in which modern industrial and financial corporations engage. While the sheer size of the capital requirements of some firms would militate against control by even their largest stockholders, the critical difference between earlier forms of capital organization and modern ones is systemic: industrial firms are no longer free to operate independently. Thus, privileged families may no longer pursue their own interests independent of the larger structure. In effect, power within corporate systems themselves has shifted away from specific corporations and those connected with them and has become a property of the larger structure in which both individual firms and privileged groups are embedded. Despite the fact that *some* families continue to wield considerable power within the financial system, no one person or family is free to operate outside of it. Each transformation in the structure of capital mobilization, consequently, has led to a decrease in the independent importance of personal ties and kinship and to a commensurate increase in the functional power of institutions and institutional arrangements within it.[112]

According to Berkowitz, the "forms" which these privileged groups assume during various stages in their transformation from small isolated clusters into larger integrated structures can be graphically represented by networks such as those shown in Figure 3.27.

Figure 3.27

Stages in the Formation of Upper Classes

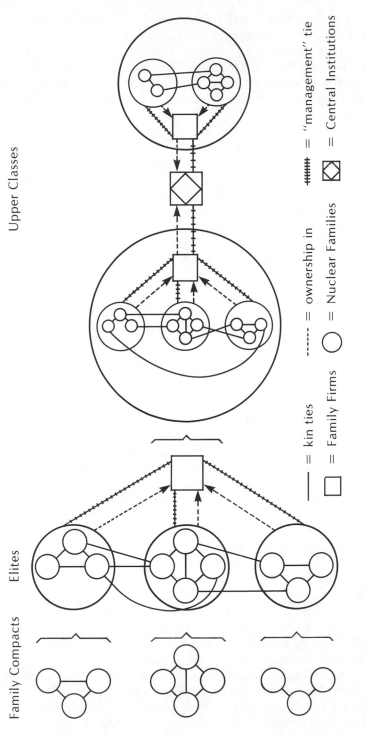

Source: S. D. Berkowitz, "The Dynamics of Elite Structure," p. 217 (modified).

Each of the small circles shown in Figure 3.27 corresponds to a nuclear family unit. The circles drawn around groups of elements indicate that at later stages in the process they have been subsumed within larger units. The brackets represent a mapping operation between layers or levels of the system. Thus, for instance, the small triad of nuclear families that appears in the upper left-hand corner of the diagram is a family compact—or group of associated merchant families—which becomes an *element* within the elite-system that appears to its immediate right. This elite, in turn, forms one element within the upper class structure. Note that the assumption here is that while kinship ties among nuclear families become more dense at different stages in the process, the boundaries between them remain unchanged. This pattern would obtain only where all systemic transformations took place within one generation. In any "real world" case, of course, this would not be true. We have made this assumption here, however, so that changes in the organization of the larger system itself will stand out. Also note that as smaller units merge into larger ones, new types of ties (e.g., ownership and management ties between family compacts and firms in elite-systems) are drawn between these larger units themselves and not their constituent parts. This is intended to show that these smaller units, for all intents and purposes, function together.

The intent of Berkowitz's and Bertaux's analogy between the structuration of upper classes and different combinations and states of physical matter is now clear: at each stage, they contend, new layers of organization and new forms of interaction between these layers may be uncovered. These "layers" generate qualitatively different consequences for the behavior of elements at each level of the structure.[113] In effect, both Berkowitz and Bertaux view the structuration of privileged groups as a process in which seemingly gradual changes in the patterns of association between institutions and family units "suddenly" become manifest in new mechanisms for capital holding and new forms of family organization. The transformation of a set of relatively isolated "molecular" family units into a systemic upper class is thus *neither* "unpredictable" and "sudden," in the conventional sense, nor simply the result of the aggregate effects of small-scale processes. It arises, in Thom's sense, from "catastrophic" properties of the *social* organization of capital and from attendant topological changes in the "social space" in which they occur.

Once again, then, the "growing edge" of structural analysis appears to be leading toward a concern with (a) the boundaries between layers within complex structures, (b) the interaction between these layers and their effects on one another, and (c) the topological properties of the spaces within which structural transformations take place. While the relationship between corporate organization and systems of privilege or class is incompletely understood, then, further research into various aspects of this problem may shed some light on these broader questions as well.

SUMMARY

During the last decade, structural analysts have been able to begin describing and analyzing aspects of the role and function of corporations in modern capitalist societies which were and are largely disregarded by conventional social scientists. In contrast to the reigning "Structure-Conduct-Performance" paradigm in economics, structural analysis has been principally concerned with the systemic relationships between and among the elementary units of corporate systems. Historically, these were poorly defined. This was not immediately apparent to most economists because the models they applied did not require them to reconcile conflicting definitions of units of analysis.

Early attempts to interpret intercorporate connections were based on simple extensions of aggregative economic models. They foundered for two reasons. First, researchers often tried to draw inferences about corporate structure, per se, from patterns of ties between and among men. Structuralists showed that there is no necessary correspondence between networks of man-to-man ties and the corporation-to-corporation connections which these analysts were implicitly interested in studying. Second, the simple ties between corporations are a poor measure of their relative "closeness" or adjacency. Structural, as opposed to nonstructural, models had to be devised which could deal with these kinds of "size" effects. Beginning around 1970, structural analysts began to create models which took these factors into account.

Initially, structuralists tried simply to map or represent complex structures of intercorporate ties. Levine, for instance, successfully used a combination of "unfolding" techniques and multidimensional scaling to create "smallest space" mappings of the directorship ties between a large number of banks and industrial corporations in the United States. The idiographic models he created, while only descriptive, overcame most of the difficulties that arise when nonstructuralist techniques are applied to these problems. More theoretically oriented algebraic techniques, such as the Lorrain-White method, can also be applied to these same kinds of substantive problems.

In the mid-1970s, the Toronto and Stony Brook groups began assembling large-scale databases on intercorporate structure. Preliminary results indicated that simple patterns of interlocking directorates are, in and of themselves, intractable. Earlier researchers had not discovered this because they had examined small samples and egocentric networks defined on the largest corporations. When systematic large-scale samples were collected, however, apparently separate networks of interlocked corporations merged into a few completely connected components within larger nets. The size of these components, moreover, was consistent with expectations derived from random models of graphs. This strongly suggested that the relatively informal clustering methods used in many earlier studies would yield inconsistent results when applied to more complete networks.

The Toronto and Stony Brook groups approached this problem in different ways consistent with their theoretical orientations. The members of the Toronto project had been strongly influenced by the exchange theory/social organization framework, which stressed the extent to which routine processes of interaction tend to produce dependency and, hence, asymmetric relationships between the elements of a system. Thus, they became interested in the ways in which various forms of interaction between firms would lead to the formation of cliques or clusters of corporations functioning, for all intents and purposes, in concert with one another. Their research, as a result, focused on discovering means of detecting enterprises, or sets of firms operating under common control, within a larger network of corporate ties.

In contrast, the Stony Brook group was explicitly seeking to establish the relative role of financial or fiduciary institutions and industrial corporations within advanced capitalist systems. This orientation derived from earlier work by Hilferding and others on systems of "finance capital." The principal tool used by this group was a centrality measure modified from one originally devised by Bonacich. By applying this measure to the patterns of interlocking directorates among a large sample of firms, the Stony Brook group was able to determine the direct *and* indirect effects of ties on given firms. Their research centered around comparing the scores of financial and industrial firms in terms of this index.

Both of these projects, and the wings of structural analysis associated with them, succeeded in developing models that could be used in interpreting corporate behavior. However, neither was able to develop methods of predicting behavioral outcomes directly from measures of the morphology of corporate structures.

Some structural analysts, however, had begun to address these same kinds of problems of inference between system levels in less formal ways. They were interested in the relationship between economic structure and class or privilege. At least two of these writers tried to look at the structuration of upper classes using elementary graph theory. Their approach to the problem strongly suggests that topological features of the social organization of capital might be important in more formal modeling of the transformation of class structures of this kind.

In the chapter that follows we will look at a number of other large-scale social processes that structural analysts have used as "experimental animals" for testing out their theories and techniques. As we will see, the problems that arise in dealing with them are different in some important respects from those we have encountered in previous chapters.

EXERCISES*

1. Using one of the standard business sources found in most libraries (e.g., *Standard and Poor's*, *Moody's*, the *Directory of Directors*), locate the members of a single corporation's board of directors. Using the "person" listings in this directory or a biographical source (e.g., *Who's Who*), create a reticulum (a network showing all the t^1 ties beginning with ego) defined on this firm. Then, go back to your original business source and locate the members of the boards of the "new" firms you found in constructing your reticulum and find the boards "they" serve on, either (a) disregarding firms that fall outside your original sample (i.e., a procedure which is comparable to Bott's or Laumann's) or (b) incorporating all new firms you may discover. Draw graphs of the reticulum and the "closed" or "expanded" reticulum. (HINT: pick an initial corporation whose board is reasonably small.) By inspection, create a narrative description to "explain" the patterns you have discovered.

2. Locate the members of a single corporation's board and create a reticulum, as in 1, above. Draw a graph of it. Then, create a graph showing its dual, following the procedures described in this chapter. Create a "closed" reticulum, as in 1, above, and construct adjacency matrices corresponding to both this reticulum and its dual. Using a piece of graph paper, "begin" plotting one of these matrices onto a graph which represents the distances between them. What problems do you encounter? (If you have access to a smallest space program, it may be possible for you to create a plot of these matrices using it.)

3. Recreate the first of the "two deans" problems outlined in this chapter. In reconstructing the elements shown in Figure 3.12, be sure to show the steps in your multiplication. (This example is simple enough to be done by hand.) Write a brief description of the composition table given in Figure 3.13. Outline, in general terms, how the reduction shown in Figure 3.14 comes about. Why does this reduction "work"?

---

* I am grateful to my 15-year-old son, Shawn Berkowitz, for testing out the mathematical aspects of the exercises for this chapter.

4. Begin with one of the standard business sources or a detailed case study of a firm. Create a table showing all of the relationships between it and other firms (e.g., directorship, ownership, debt relations, joint production agreements, joint ventures, shared technology agreements). For each such alter, specify whether it is (a) a subsidiary, (b) a jointly held firm, (c) a "partner," (d) a supplier, (e) a firm supplying goods or services to ego, (f) a lender of capital, (g) a borrower of capital, or (h) a firm enjoying some other type of relationship to the initial firm. Define a reticulum for each of these types of ties, and superpose them. To what extent do they map over one another? How are they different? Specify a set of mathematically consistent operations you could perform which would bring out these differences.

ADDITIONAL SOURCES

Burt, Ronald S. "Cooptive Corporate Actor Networks: A Reconsideration of Interlocking Directorates Involving American Manufacturing." *Administrative Science Quarterly*, forthcoming.

Coleman, James S. "Clustering in N Dimensions by Use of a System of Forces." *Journal of Mathematical Sociology* 1 (1970): 1–47.

Coleman, James S. *The Mathematics of Collective Action*. Chicago: Aldine, 1973.

Davis, James A. "Clustering and Hierarchy in Interpersonal Relations: Testing Two Graph Theoretical Models on 742 Sociograms." *American Sociological Review* 35 (1970): 843–52.

Erickson, B. H. "Some Problems of Inference from Chain Data." In Karl Schuessler (ed.) *Sociological Methodology, 1979*. San Francisco: Jossey-Bass, 1979, pp. 276–302.

Evan, W. M. "An Organization-Set Model of Interorganizational Relations." In M.F. Tuite, M. Radnor, and R. R. Chisholm (eds.), *Interorganizational Decision Making*. Chicago: Aldine, 1972, pp. 181–200.

Frank, Ove. "Sampling and Estimation in Large Social Networks." *Social Networks* 1 (1978): 91–101.

4

Community-Elite Networks and Markets: Structural Models of Large-scale Processes

THE SETTING

Any science, it has been observed, ultimately owes as much to its "experimental animals" as to almost anything else. Modern genetics, for instance, would probably have progressed much more slowly without *Drosophila melanogaster*, the common fruit fly, whose rapid breeding and large chromosomes proved so useful to studies of inheritance.[1] Archaeologists owe a similar debt of gratitude to the lowly potsherd, or piece of earthenware pot, whose convenient habit of depositing itself in layers has enabled scholars to accurately date much more prepossessing artifacts and relics. Physicists have learned a great deal from two "children's toys"—the pendulum and the prism—and from what was often considered a "waste product"—pitchblende. In the same vein, anthropologists have favorite "tribes," and survey samplers create "panels" of respondents whose views on everything from nuclear disarmament to designs for automobile grills are periodically collected and assessed.[2]

In this chapter we will briefly look at two types of "experimental animals,"—*community-elite networks* and *markets*—which have provided especially good testing grounds for structural analytic ideas and methods. Just as other scientists choose the creatures or materials they observe because of traits which make them unusually well suited to a particular purpose, structural analysts have good reasons for being interested in these specific structures. First, both community-elite networks and markets have been intensely examined by other social scientists. In practical terms this is useful because it means that the results obtained by studying them in a structuralist fashion can be compared to results arrived at by more conventional means.[3] Second, standard methods of dealing with both subjects leave many important questions unanswered. By showing that they are able to handle some of these issues, structural analysts can provide a clear demonstration of the value of both their modes of conceptualizing problems and their techniques. Finally, community-elite networks and markets have historically been treated psychologistically. By dealing with them sociologistically, structuralists place themselves in a good position to discover new problems implicit in these "experimental animals" which other researchers have overlooked. In the long run, these "new" questions may be more important than many of those which either conventional or structuralist social scientists can identify at the moment.[4]

Thus, while structural analysis as a whole has paid a good deal of attention to the issues we discussed in the last two chapters, it has not dealt with other specific interpersonal or corporate structures as intensely as those we will examine here. This is, of course, one of the best reasons for having standardized "experimental animals" in the first place: they allow researchers to focus on more circumscribed and, hence, more clearly delineated problems.

CASE I: THE COMMUNITY-POWER CONTROVERSY

Throughout the 1950s, 1960s and early 1970s, social scientists from a number of disciplines were involved in a heated debate over the bases of political and social power within communities. This debate was, in turn, part of a larger conflict between *pluralists* and *antipluralists* which centered around the extent to which corporations and business interests are able to dominate the decision-making processes in advanced capitalist societies.

The pluralists held that *power* in Western democracies is vested in governments which are subject to cross-pressures or influences from a series of competing and often divergent "interest groups."[5] While these groups can (and often do) form coalitions in order to effect changes in policy or shape legislation, alliances between and among them are, of necessity, transient or at least impermanent. Thus, pluralists argued, the institutionalized *divisions* between these groups ensure that, in the long run, no interest or set of persons vying for power will predominate.[6]

By contrast, antipluralists maintained (a) that these "interest groups" are really economic and political elites who represent no interests save their own; (b) that some elites play a central role in the formation of public policy, while others are quite peripheral to it; (c) that, on balance, economic elites and economic interests are dominant; and (d) that this predominance of economic elites reflects the class structure and hence class basis of political power in advanced capitalist societies.[7] "Competition" between sets of persons vying for power thus reflects slight differences or disagreements between "factions" within a single, unified *ruling class*.[8]

Beginning in the early 1950s, social scientists supporting these positions conducted a series of studies of community decision making which were designed to "test" their models empirically. Typically, one of three kinds of strategies was employed. Pluralists tended to focus on concrete issues facing communities and to use participation in these issues as a touchstone for uncovering "influentials" who had helped to shape actual decisions.[9] Antipluralists, by contrast, usually utilized what was called the *reputational method* in which respondents were asked to identify influentials directly or simply to list people "who counted" within a given community.[10] In some cases both pluralists and antipluralists made use of a technique in which researchers (often with the help of respondents) identified "positions" which "make a difference" in community affairs (e.g., president of a local bank, chairman of the school board). Persons occupying these positions were then designated as "influentials."[11]

Each of these strategies, in and of itself, is seriously flawed. The *community issue method*, as Robert Perrucci and Marc Pilisuk point out, does not account for the possibility that

> power may be at work in maintaining the directions of current policy, in guarding the agenda of public controversy so as to preclude [them] from reaching the status of a community issue. Such guarding of the agenda could be achieved, for example, by domination of the media, by political patronage, and by effective public relations and lobbying.[12]

The reputational method, while likely to uncover the less-public centers of power involved in shaping policy, has been justly criticized by M. Herbert Danzger and others on the grounds that having a reputation as a power-holder and actually exercising power are not the same thing. By asking respondents to name those who "count," researchers simply feed into respondents' biases about the issues which are most important to the life of a given community. By contrast, Danzger argues for methods which treat the "issue areas" used in measuring community power as problematic.[13]

Positional methods, while more likely to detect those exercising power or influence "behind the scenes" than the community issue technique—and less likely merely to reaffirm researchers' or respondents' biases than the reputational method—ultimately rest on an *institutional* theory of power, that is, that an individual's "power" derives from his or her location within an organizational nexus. By simply using "positions" or institutional locations as a means of locating "influentials," it is argued, researchers remove the context necessary to interpret actors' behavior, that is, the institutional "constraints" under which they operate.

By the mid-1960s, many of these methodological inadequacies were recognized. John Walton, in a key study published in 1966, reviewed the evidence presented in 33 studies of 55 communities. His chief finding, as we might suspect, was that the methods used by pluralist and antipluralist researchers largely determined the particular substantive results they obtained. Clearly what was needed was some set of techniques which would allow researchers to examine these same issues, but in a more dispassionate fashion.[14]

Community-Elite Networks

In 1963, a group of structuralists—Linton Freeman, Thomas Fararo, Warner Bloomberg, and Morris Sunshine—explicitly compared a number of different methods of "locating" community leaders. Among other things, they found that the "community influentials" identified in reputational samples tended to head organizations whose members were the most active in community affairs.[15] This strongly suggested that there is a relationship between the number of ties formed by members of an organization and the repute in which its members are held within a given community. Freeman et al. verified this by creating a measure of "organizational participation" which could be associated with individual rankings on reputational measures.[16]

In 1968, Freeman followed up this work by directly examining organizations whose members had participated in community decisions. Substantively, he found that clusters of these organizations did not closely correspond to those which would have been predicted by either pluralist or antipluralist models; that is, they were quite heterogeneous. This was an extremely useful finding because it demonstrated empirically what a number of researchers had suspected: both the classic pluralist and antipluralist positions were far too simplistic and needed to be restated in more sophisticated forms.[17]

Methodologically, Freeman's study accomplished something which in the long term was probably even more important: it helped to move the focus of the "community power controversy" away from individual *men*—their power, influence, and so forth—and toward an examination of systems of linkages between and among *organizations* or *groups*. In effect, the "community influentials" issue had been transformed into the "community-elite network" question.

In the same fashion that we noted in the last chapter, however, most researchers did not immediately recognize the larger implications of this shift in emphasis from man-to-man to organization-to-organization or group-to-group networks. As late as 1970, for instance, Perrucci and Pilisuk attempted to draw inferences about "interorganizational links" and "resource networks" from graphs in which organizations are represented as nodes and men as relations between them.[18] Predictably, however, their analysis focused on the characteristics or attributes of linkages (men) rather than the morphological or structural properties of these networks themselves, that is, on the system of organization-to-organization linkage about which they were most interested initially. Their final analysis, as a result, was based on the dual of their original datagraph.

Structural Models of Community-Elites

As we mentioned earlier, the most developed and sophisticated structural studies of community-elite networks have been conducted by Laumann, Galaskiewicz, Marsden, and Pappi, and were published in a series of joint papers and monographs beginning in the early 1970s. Here we will initially focus on one study reported by Laumann and his colleagues which succinctly illustrates how proximity measures and smallest space analysis can be used to examine the kinds of issues raised by pluralists and antipluralists a decade ago. We will then look at Breiger's reanalysis of these same data with an advanced algebraic technique called blockmodeling.

In this case, our "experimental animal" is a dataset produced through a unique combination of community-issue, reputational, and positional techniques in the city of "Altneustadt," West Germany.[19] Initially, the principal investigators during this stage of the research—Laumann and Pappi—identified a list of community leaders using positional methods. Persons on this list were then asked to name "reputational leaders." Five issues were selected and used to gauge the degree of consensus or disagreement among members of this elite. Then, through a series of personal interviews, data were gathered on leaders' reactions to these issues, on their backgrounds, and on three separate measures of contact among them. Forty-six of 51 community leaders included on the combined lists were eventually interviewed.[20]

In the study we are principally concerned with here—"The Analysis of Oppositional Structures in Political Elites: Identifying Collective Actors"—Laumann and Marsden were interested in "oppositional structures" within elite decision-making systems.[21] They begin by defining units of analysis which they refer to as *collective actors*:

> A *collective actor* is defined as a subset of members in an elite decision-making system that includes only those (1) who share an outcome preference in some matter of common concern, *and* (2) who are in an effective communication network with one another.[22]

Each collective actor, defined in this way, consists of a cluster of individuals (nodes) (a) sharing, but not necessarily acting in terms of, a set of common orientations toward an issue, (b) in which each member is some specifiable finite path-distance from the others. Collective actors are graphically represented in Figure 4.1.

Figure 4.1

Collective Actors as Defined by Laumann and Marsden

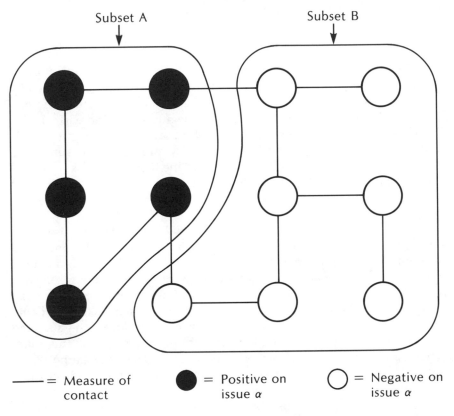

Subset A   Subset B

——— = Measure of contact   ● = Positive on issue α ○ = Negative on issue α

Note that all of the nodes represented in Figure 4.1 are connected to one another at some distance $t \leq t^N$. The members of the two collective actors shown here (subsets A and B), however, *also* share a property (or "color") which discriminates between them.[23] Therefore, while the set of nodes given in Figure 4.1 *as a whole* meets one of the conditions given in Laumann and Marsden's definition (condition (2)), it does not meet both. Further note, however, that neither subset A nor B constitutes a clique in the strict sense: the definition given above does not require elements within collective actors to be adjacent to one another, but simply reachable at some distance.[24]

In order to deal with these units analytically, Laumann and

Marsden reduce each collective actor to a single point by transitively closing a digraph defined onto their constituent nodes, that is, by reducing the path-distances between vertices to zero. In this case, these path-distances are calculated by combining three measures of "effective communication" between elites: (a) "social contact," (b) "business relationships," and (c) "discusses community affairs with." Since, in the interviews, respondents were asked to name the three persons to whom they were most closely connected in each of these ways, Laumann and Marsden adjust this distance measure so that (a) each cluster (collective actor) contains a minimum of three nodes, and (b) all collective actors are mutually exclusive with respect to a given issue.[25]

Having reduced each collective actor to a single point, the authors then proceed to calculate a measure of the interpoint distance between them. Since, by definition, collective actors are mutually exclusive with respect to any given issue, this is done by comparing their pair-wise similarities and dissimilarities across all five issues about which respondents were questioned during the interviews.[26] The similarity/dissimilarity measure obtained in this way is then used to map collective actors into a smallest space model following the procedures we described in earlier chapters.

Laumann and Marsden argue that there are four theoretically germane configurations which could result from a procedure of this kind, each of which corresponds to one of the models proposed during the community power structure controversy. These configurations are depicted in Figure 4.2.

Each of the letters shown here corresponds to a collective actor. The type of letter (A, B, C, D, or E) indicates the issue with which that collective actor was concerned. The subscripts indicate whether it was an opponent (o) of the policy designated by the letter, or a proponent (p) of it. In Figure 4.2a, Laumann and Marsden present a model which, they argue, represents the "oppositional structure" one would expect in "a classic . . . pyramidal . . . or elitist structure."[27] Note that in this "centrally administered" model "all winning collective actors are coincident with one another . . . and there exist no collective actors opposing their dominion over the community." Thus, while some of the collective actors depicted in Figure 4.2a oppose a policy and some support it, "there is a complete absence of any losing collective actors."[28] In Figure 4.2b, Laumann and Marsden seek to represent "the classic polarized, bifurcated community leadership system." This is the sort of structure we might expect if some strong basic issue, such as religion, effectively divided a community-elite into two "factions."[29] Note that, in this model, the clusters appear at the extreme left and extreme right of the space. In the limiting case, this would reflect the fact that they were as far away from one another as measurement would

Figure 4.2

4.2a: Centrally administered model, no oppositional collective actors

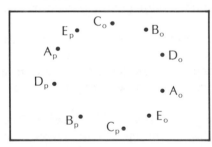

4.2b: Unidimensional oppositional model, recurrently opposed collective actors

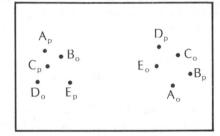

4.2c: Multidimensional cleavage model, coordinating center absent

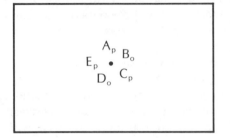

4.2d: Polycentric bargaining model, with widely scattered opposed collective actors

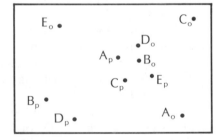

Source: Edward O. Laumann and Peter V. Marsden, "The Analysis of Oppositional Structures in Political Elites" (1978), p. 127. Reprinted by permission.

permit. Laumann and Marsden's third model, which appears in Figure 4.2c, would represent one of the situations implicit in the pluralist position: the case in which each special interest group was narrowly focused around its own concerns and did *not* engage in bargaining or coalition formation. Collective actors in this case "tend to be relatively homogeneous in their composition and are likely to pursue narrowly defined special interests."[30] Finally, in Figure 4.2d, Laumann and Marsden present an example of what they refer to as "a polycentric bargaining structure" in which "at least one contesting collective actor for each issue is composed of individuals recruited from quite diverse institutional arenas."[31] These tend to occupy the core, while less diverse "challenger" groups are on the periphery. This model, in effect, is what we would expect to find if the classic pluralist position were sustained;

that is, it represents the maximum opportunity for bargaining, coalition formation, and so forth.[32]

Laumann and Marsden then compare these ideal-type models with the smallest space model they are able to generate from their data. This is shown in Figure 4.3.

Note in Figure 4.3 that "winners" (indicated by boxes) in the actual decisions arrived at in "Altneustadt" appear to the center-right of the diagram and "losers" to the center-left. Laumann and Marsden explain that, in "Altneustadt"

> The "losers-winners" axis . . . follows the recurrent ideological split between the socially and economically liberal or conservative factions that participate in various community controversies. The second axis at almost a right angle to the first simply distinguishes between two pairs of collective actors, the first of which (G_{o_1} and G_{o_2}) share opposition to the construction of a new city hall and the second of which (M_{o_1} and M_{o_2}) share opposition to the pop musical festival. Since each pair lacked common communication links, they had to be located at considerable distance from one another. . . .G_{o_1} is a collective actor consisting of S.P.D. (German Socialist Party) supporters and fellow travellers . . . who hold highly egalitarian social views, while G_{o_2} are economically conservative C.D.U. businessmen with economic interests in the county (rather than the city) who strongly objected to the building of an expensive city hall by their usual allies.[33]

Laumann and Marsden conclude that, in general form, the oppositional form shown in the "Altneustadt" data most closely approximates the hypothetical model given in Figure 4.2b: the "recurringly opposed" factional elite case.

These results generally accord with those Breiger arrived at by using algebraic modeling techniques. The method he applied, *blockmodeling*, is, as we mentioned before, an extension of the logic of the Lorrain-White method, and was developed by Boorman, Breiger, Heil, White, and others in the early 1970s.

Blockmodeling of Social Structures

A blockmodel is a network which is a structural abstraction of a data network.[34] In a blockmodel, single "model nodes," called *blocs*, are used to represent *sets* of nodes in a *data network*. Ties between *data nodes* are represented by ties between corresponding *blocs* in the blockmodel. Thus, if a data network has c types of ties, represented by c graphs or matrices, on the same p nodes, then the model consists of an "image," with c types of ties on b blocs ($b \leq p$). Consequently, a blockmodel is valid if and only if there is some mapping of the p nodes into the b blocs such that the ties in the *image graph* adequately represent those in the data network.[35]

Figure 4.3

Two-dimensional Configuration of the Oppositional Structure of Collective Actors in "Altneustadt"

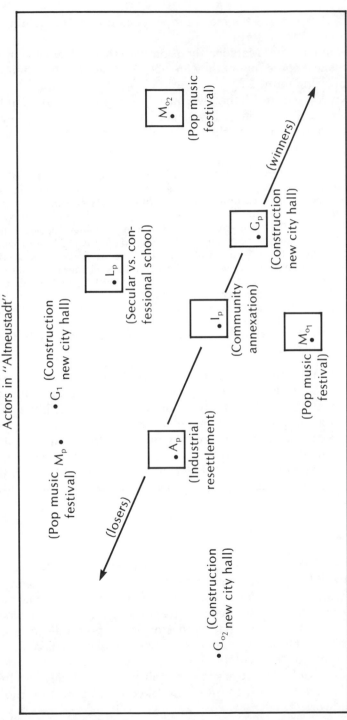

Source: Edward O. Laumann and Peter V. Marsden, "The Analysis of Oppositional Structures in Political Elites" (1978), p. 160. Reprinted by permission.

For instance, let us assume that we have the same data we had in our simplest example of the deans' office problem (Figure 3.19). These can be arranged into a square matrix, as in Figure 4.4.

Figure 4.4

	1	2	3	4	5	6	7
1	0	1	1	0	0	0	0
2	0	0	0	1	1	0	0
3	0	0	0	0	0	1	1
4	0	0	0	0	0	0	0
5	0	0	0	0	0	0	0
6	0	0	0	0	0	0	0
7	0	0	0	0	0	0	0

Here we see the pattern of ties for what was referred to earlier as the C, or "command" relationship. In this case, the nodes are labeled so as to avoid confusion later on. If we *hypothesize* that there are three *roles* which these nodes may play—that is, *pure senders*, who send but do not receive instructions; *sender/receivers*, who both send and receive commands; and *pure receivers*, who only receive them—we can represent this hypothesis in the form of an image graph, as in Figure 4.5.

Figure 4.5

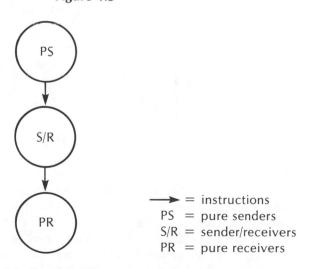

⟶ = instructions
PS = pure senders
S/R = sender/receivers
PR = pure receivers

This graph corresponds to the binary-image matrix given in Figure 4.6. Note that in Figure 4.6 ones in this matrix correspond to ties between blocs, or sets of nodes, rather than individual offices. Similarly, zero-entries indicate that, hypothetically, no such ties exist. These hypotheses can be juxtaposed to data by simply *partitioning* the original data matrix into submatrices, or *blocks*, as in Figure 4.7.

Figure 4.6

	1	2	3
1	0	1	0
2	0	0	1
3	0	0	0

Figure 4.7

		Bloc 1	Bloc 2		Bloc 3			
		1	2	3	4	5	6	7
Bloc 1	1	0	1	1	0	0	0	0
Bloc 2	2	0	0	0	1	1	0	0
	3	0	0	0	0	0	1	1
	4	0	0	0	0	0	0	0
	5	0	0	0	0	0	0	0
Bloc 3	6	0	0	0	0	0	0	0
	7	0	0	0	0	0	0	0

Note that this three-bloc partition generates nine submatrices showing ties from one of the three sets of officeholders to the others. Each block which is predominantly zero-filled is what we refer to as a *zero-block*. *One-blocks*, by contrast, contain ones. In constructing blockmodels we demand that "0" ties in an image matrix correspond to zero-blocks and that "1" valued ties correspond to one-blocks in a partitioned data matrix. Under certain circumstances, however, one-blocks and zero-blocks need not be "pure," that is, they may contain some zeros or some ones, respectively.[36] As a rule, strategies of blockmodeling emphasize the importance of zero-blocks in constructing and validating hypotheses.

While the precise reasons for this are too technical for us to go into in detail here, we can see intuitively that one of the weaknesses in the Lorrain-White method (and, hence, one of the reasons that blockmodeling was developed) is that it *demands* that ties be made. Thus, for instance, as far as the Lorrain-White method is concerned, a hierarchy that is "missing" only one tie is *not* a hierarchy; that is, the semi-group that it generated would be quite different from that produced by a pure hierarchical pattern. Blockmodeling is more "permissive," and hence more adaptable to the "real world."

In effect, then, blockmodeling most frequently uses the *holes* in structures (i.e., the places where they are not connected) rather than their points of positive connection as a way of describing them. Thus, in Figure 4.7, the three zero-blocks corresponding to the first column of the matrix verify that the most significant aspect of our definition of pure senders has been met: bloc 1 (node 1) does not receive ties from another bloc. It is not isolated, however, because it does send ties to bloc 2 (nodes 2 and 3). Similarly, bloc 2 receives ties only from bloc 1 and sends them to bloc 3: that is, it meets the essential features of the sender/receiver role, as defined. Bloc 3 (nodes 4, 5, 6, and 7), as we hypothesized, is a pure receiver.[37]

Note that here, as in the Lorrain-White method, all of these hypotheses can—and, in fact, must—be tested simultaneously. Further note that other important algebraic properties of Lorrain-White models obtain as well: by simultaneously mapping both the nodes and relations in the data graph into partitions corresponding to nodes and relations in the image graph, we achieve the same kind of homomorphic reduction as we achieved earlier. Each bloc, moreover, is homogeneous with respect to both its *internal* and *external* ties.[38] Blockmodeling, therefore, also deals simultaneously with both the general *roles* played, and particular *positions* of, nodes within an overarching structure.

Under normal circumstances, of course, we would not apply blockmodeling to a case as simple as the one we have here. Moreover, the real strength of blockmodeling as a technique lies—as it does in the Lorrain-White method—in its ability to deal with more than one type of relation defined onto the same nodes. This can be accomplished in much the same way that it was in our example in Chapter 3: by distinguishing different *types* or forms of relationships from one another, creating *compound images* reflecting combinations of these, forming the multiplication table showing interlocking roles, structures, and so forth. When the "blockings" produced in this fashion are nontrivial and consistent *across* a set of networks, and when they generate consistent *patterns* in data matrices, blockmodels should reflect the underlying logic or structure in the "real world" situation they are being used to model.[39]

Breiger on "Altneustadt"

Breiger's reanalysis of the "Altneustadt" data clearly shows how this is done. He begins by translating a series of models of community-power structures, taken from research done by Peter Rossi, James Coleman, and others,[40] into a set of image matrices. These are shown in Figure 4.8.

Figure 4.8

Image Matrices Reflecting Different Conceptions of Community Power

4.8a: Caucus Rule   **4.8b:** Multiple Caucus

1	0
0	0

1	0
0	1

4.8c: Hierarchy and Deference   **4.8d:** Center-Periphery

1	0	1	0
1	0	1	1

1	1
1	0

4.8e: Amorphous

1	1
1	1

Source: Breiger, "Toward an Operational Theory of Community Elite Structure," *Quality and Quantity* 13 (1979): 21–47 at 27–31. Reprinted by permission of the publisher.

In each case, "the elite members who are active in the network [of key social ties within the community] comprise the first block, and the others (who may be active in *other* networks) comprise the second."[41] In the "caucus rule" model, a compact elite largely monopolizes "both the initiation and reception of social ties of the relevant type."[42] Decisions are made within this group, and participation by "outsiders" is discouraged. The "multiple caucus rule" model is a generalization or refinement of this, in which it is possible for *different* elites to interact with one another—each, in effect, specializing in a specific policy area—and for cleavages and cliques to occur within one elite. The "hierarchy" and "deference" models reflect Breiger's interpretation of two structures implied in Peter Rossi's "pyramidal" model. Both try to capture structures in which there are clear strata, either within an elite or "across the population."[43] In the first, all ties go *to* the "leadership bloc." In the second, the lower stratum is internally connected, but "defers" (is tied to) the leading group as well. The "center-periphery model" reflects a situation in which "a coherent set of active members (or a 'leading crowd') is surrounded by isolated individuals who have interchange both *to* and *from* them." The "amorphous" model represents a residual form.[44]

Breiger then distinguishes among the measures of contact between leaders which we mentioned earlier, and constructs a single matrix for each set of ties. He designates these as "BP" (business/professional relations), "COM" ("discuss community affairs with"), and "SOC" (social contacts). Using the cliques identified earlier by Laumann and Pappi—the CDU clique, the county businessmen's clique, the SPD clique, and a residual category (which Breiger calls *bridgers*)—he then partitions 44 of the leaders into a series of sets and applies these groupings to each of the three matrices of relations.[45] Since, in real applications, blocks are seldom "pure," he then uses a density measure to discriminate between zero-blocks and one-blocks (which he calls *bonds*).[46] The data matrices produced by this "cleaning" procedure are given in Figure 4.9.

Figure 4.9

	BP	COM	SOC
CDU	110	100	100
BUS	100	100	010
SPD	001	001	001

Source: Breiger, "Toward an Operational Theory of Community Elite Structure," *Quality and Quantity* 13 (1979): 21–47 at 37. Reprinted by permission of the publisher.

In each of these three matrices, the CDU and SPD cliques are internally coherent (they are 1-tied to themselves), while the county businessmen (BUS) are only strongly tied to themselves socially. In other respects, the businessmen are structurally subordinated to the CDU clique. This is consistent with earlier observations made by Laumann and his colleagues. In addition, however, Breiger is able (a) to provide a detailed picture of the relationships between the CDU and county businessmen's cliques because of his ability to separate out the effects of different types of ties on the structure, and (b) to describe the patterns of reciprocity or dominance among the cliques within each of the three domains.

The strengths of blockmodeling, however, only become clear when Breiger brings the more amorphous bridgers into the picture. Versions of the data matrices shown in Figure 4.9 which include this group are given in Figure 4.10.

Figure 4.10

CDU	1 1 0 0	1 0 0 0	1 0 0 0
BUS	1 0 0 0	1 0 0 0	0 1 0 0
SPD	0 0 1 1	0 0 1 0	0 0 1 1
BRID	0 1 1 0	1 1 1 0	1 1 0 0
	BP	COM	SOC

Source: Breiger, "Toward an Operational Theory of Community Elite Structure," *Quality and Quantity* 13 (1979): 21–47 at 36. Reprinted by permission of the publisher.

Here we can see that while the SPD clique has directed social contacts with the bridgers, these are not reciprocated. Since respondents had been asked to name the three persons with whom they had the *most* social contact within the leadership group, this reflects the apparently different value attached to these social encounters by members of these two groups.

Breiger notes, moreover, that a clear hierarchical pattern obtains in the pair-wise "community affairs" contacts, between the SPD and bridgers, BUS and CDU, and CDU and bridgers cliques. These would accord with the "hierarchical" image matrix postulated earlier:

$$\begin{bmatrix} 1 & 0 \\ 1 & 0 \end{bmatrix}$$

Breiger's analysis of the business/professional relations matrix, however, is probably the most interesting from our point of view. Here he notes that apart from the tie between the bridgers and the county businessmen's group, the structure divides neatly into two caucuses: one including the county businessmen, but headed by the CDU clique, and the other led by the SPD, but including the bridgers.[47] This finding directly corresponds to the oppositional structure discovered by Laumann and his colleagues and shown in Figure 4.3.

In the rest of this paper, Breiger is able to refine each of these observations further and to construct a compound role structure reflecting the interlocking between each of these forms of contact among members of the "Altneustadt" leadership group. In the course of this, he establishes that while each of the separate contact matrices shows evidence of the "hierarchy" and "deference" patterns depicted in Figure 4.8, the role structure of the group as a whole is more complex. On balance, it most closely approximates the "multiple caucus" model. This pattern, however, is overlaid by another in which there is evident inequality in the relations *among* these caucuses.[48]

Structural Models and Community Power

Both of these structural methods, then, have allowed researchers to examine far more subtle and detailed aspects of the organization and exercise of power within communities than those which preoccupied pluralist and antipluralist writers a few years ago. By transforming the "community influentials" problem into the "community-elite network" question, structuralists were able to shift much of the debate away from issues which in the final analysis are probably not resolvable in their own terms and toward more measureable features of the institutional landscape. In the process, they discovered a number of interesting things about the social bases of individual power and the mechanisms through which these factors come into play. Thus, while community-elite network studies obviously cannot answer some kinds of questions about the organization of power at other levels in a society, they have proved to be an extremely useful "experimental animal" for testing out theories in a more limited and hence more interpretable context.

CASE II: THE STRUCTURE OF MARKETS

By this point it should be clear that there is no cut-and-dried "structural analytic method" which can simply be applied to all situations. As a result, developing adequate structural models usually requires imagination and effort: few areas are so well explored that "cookbook" solutions will work. Thus, in most cases, *doing* structural analysis means being involved in a process of discovery in which new possibilities are often accompanied by knotty technical problems.

Following up these "leads" and finding solutions for these difficulties, however, is what science is all about. Thomas Kuhn observed that when a field reaches the point where the same models are being applied to marginally different events, research is increasingly likely to confirm, rather than challenge, the established wisdom. New discoveries can then only come about from radically different ways of conceptualizing old problems or from vastly improved techniques which, in themselves, make it possible for researchers to formulate new questions.[49]

As a rule, both new frameworks and new tools are necessary before a "scientific revolution" can get underway. In the previous section we saw that the community-power controversy had reached an impasse because both pluralist and antipluralist researchers thought of power in individualistic terms. By restating the problem sociologistically, structuralists had been able to deal with it in a more empirically open fashion. However, if structural analysts had not had some powerful way of exploring the implications of their reformulated version of the question, it is unlikely that their models, no matter how valid, would have been treated as anything more than mere speculation.

In this section, we will examine a case where this process of discovery is still in its formative stages. While structural analysts have taken the first steps toward redefining "markets" in systemic terms, the tools they have developed for analyzing the consequences of different market-structures are still relatively primitive. Thus, the "scientific revolution" in this area is not yet complete. However, by dealing with a problem in this stage of its history, we should be able to examine the *process* of scientific discovery itself in a more detailed way than we could where theory and methods were already highly developed.

Markets and Structures

As we noted in Chapter 3, conventional economists conceive of markets as abstract arenas where "buyers" and "sellers" meet (in a figurative sense) to exchange goods and services. While, due to sheer parsimony, this notion of a market is attractive, it is also restricted in a number of important ways. First, an abstract market of this kind cannot be empirically observed because there is nothing which corresponds to it in the "real world." It is a pure *construct*. Hence, while we can directly observe *behavior*—of individuals, companies, institutions, or whatever—which can be interpreted as *market behavior*, we must *infer* the existence of markets themselves from it.[50] Second, classic market notions specify only two roles which participants may adopt—buyers or sellers. This prejudges the *types* of interaction that may go on between actors.[51] Third, since these buyer and seller roles are mutually exclusive and exhaustive, this formulation of the idea contains strong presumptions about the *forms* or patterns of interaction in which participants may engage. Finally, because conventional notions of markets are so abstract, it is often difficult to establish the range of conditions under which they apply.

While most economists recognize that models based on this classic notion of a market are not intended as perfect analogs to "real world" situations, many of the implications of these restrictions are ignored in practice. For instance, if a market cannot be directly observed, then, strictly speaking, analysts must validate or establish its existence in each instance where they employ the term—as social scientists do when they use other second-order concepts such as "class consciousness." Yet, the conventional economic literature is filled with instances where the idea of a "market" is not only assumed, but is used to validate *other* phenomena, for example, attributing a fall in the value of the dollar to "market forces."[52]

Despite problems of this kind, economists have made few attempts to rework or rethink their notions of markets or of market behavior since

Edward Chamberlain's unconventional *Theory of Monopolistic Competition* was published in 1933.[53] A. Michael Spence's *Market Signaling,* and other work following along these lines, broadened the classic notion of markets by specifying the consequences of nonprice "signaling" between participants.[54] And John Harsanyi, via game theory, has proposed alternatives to the simple bilateral bargaining implicit in most market models.[55] However, with rare exceptions, the classic notion has largely been left intact.

Structural Studies

In the late 1960s and early 1970s, a number of structuralists began to examine markets and marketlike structures. While they were dealing with substantially the same phenomena which economists construe as market behavior, they did not initially conceive of what they were doing in these terms. Nancy Howell, for instance, studied the ways in which women used contact networks in order to procure illegal abortions. However, since her focus was on the process of information flow and how participants sought to protect themselves (and the abortionist) from the consequences of information dispersal, she did not center her analysis on the relationship between the "search" process itself and the structure within which it was taking place.[56] Despite this, her work contained the germ of a structural theory of markets in that it specifically dealt with the mechanisms through which scarce goods are allocated.

However, Mark Granovetter, in a study of job searching and job changing among professional, technical, and managerial personnel in a Boston suburb, explicitly juxtaposed his findings to the models of labor markets used by economists and manpower specialists.[57] His results starkly contrast with the assumptions underlying conventional market models: "A very important part of labor-market behavior," Granovetter argues, "is . . . inadequately described by search models."[58] Instead, Granovetter suggests, in a large number of cases a "matching" process goes on in which employers and their agents recruit candidates for jobs—and what he calls quasi jobs (jobs created with a particular person in mind)—through informal contact networks. This process, he demonstrates, is best understood as a socially organized and regulated phenomenon which does not obey the price-signaling which economists see as the principal mechanism which ties supply and demand, and by extension buyers and sellers, together within markets. In fact, the highest-paid people Granovetter interviewed were the least likely to have obtained their jobs through formal channels and the arm's length transactions stipulated in classic market models. "Active searchers"— those explicitly in the labor market in the classic sense—tend to be less well paid. Thus, wage inducements (price) appear to be an infrequently used way of mobilizing supply (workers).[59]

Granovetter did not try to use his findings as a beginning point for a more general exploration of market behavior. In fact, throughout the study he draws a distinction between the structured job-getting process he is primarily interested in and more random market behavior. In effect, he did not see markets, per se, as structured, but as a kind of residual arena for job seekers who lack the contacts necessary to find employment through more structured means.[60]

In 1975, Scott Boorman published a paper, entitled "A Combinatorial Optimization Model for Transmission of Job Information through Contact Networks," which may well prove to be one of the most important in the development of structural analysis.[61] The paper accomplishes a number of things simultaneously. First, Boorman shows that both informal structures (such as Granovetter's contact networks) and more formal structures can be modeled using techniques which assume that individual actors attempt to optimize the allocation of scarce resources. This is important because it enables Boorman seriously to call into question the distinctions which social scientists have traditionally drawn between formal and informal organizations. Second, his model specifically deals with the ways in which patterns of "weak" and "strong" ties linking individuals to one another—which Granovetter had shown played very different roles in the job-search or job-changing process—can be formally understood in the same terms that economists and game theorists use to describe systems of exchange or markets.[62] This is critical because it argues, inter alia, that markets *are* structured through networks of ties reflecting concrete relationships and channels of information flow among actors. Finally, Boorman shows that expectations derived from his model closely coincide with the empirical evidence that Granovetter gathered. This would tend to validate Boorman's model, because it is tightly formulated and, hence, would be unlikely to behave in this way unless it had captured the underlying logic of the structures Granovetter had been studying.[63]

In essence, Boorman succeeds at one stroke in (a) redefining "markets" as concrete entities which correspond to potentially observable "real world" events, namely, ties between and among actors; (b) showing that the underlying logic of these ties is, itself, rational; (c) demonstrating that this redefinition of "markets" and underlying logic are consistent with the abstract idea of an economic "market"; and (d) showing that in at least one case this formulation is consistent with a set of empirical findings.

Markets as Mobilization

Boorman's model is far too elaborate and technical to be described in detail in a book of this kind. It is clear, however, that it represents a major step toward the kinds of conceptual apparatus and technical tools which are necessary if structural analysis is to provide robust proof of its ability to deal with markets and marketlike phenomena.

Two recent developments suggest that structural analysis may not be far from the point where it can do this. First, Harriet Friedmann has developed a structural model to explain the evolution of the world wheat market and its relationship to the larger dynamics of an expanding industrial capitalist economy.[64] The core of her argument is that the world wheat market as it developed at the end of the nineteenth century did not result from a series of abstract exchange relationships which were simply knit together into a larger whole. Rather, it was the result of a process in which *both* sources of supply and consumption markets were simultaneously created to fill a niche in Europe's developing industrial system.

Friedmann's argument, then, while ultimately based on historical evidence, is formally similar to Boorman's in its central premise: that markets are formed by ties among concrete social actors in response to specifiable factors and within given constraints. She adds a dynamic dimension to this model, however, by stressing the role played by these actors in forming or creating this market as a concrete entity.[65]

Second, in a recent set of working papers, Harrison C. White has begun to elaborate a formal model based upon the idea of markets as "social structures", involving concrete actors (firms) who respond to "tangible signals," terms of trade which they observe in actual markets.[66] While White uses many of the same tools neoclassical economists do in formalizing this model, it is ultimately based upon different presumptions about the nature of the "marketplace," possible forms of interaction among participants, and the conditions under which new markets will be formed. In addition, it suggests circumstances under which loosely connected market structures would be replaced by tight-knit formal organizations seeking to eliminate transfer costs.

While it would be premature to argue that these three different perspectives constitute a structural "theory" of markets, they appear to be converging toward something more than a mere series of related models. The implications of such a theory, moreover, would extend well beyond "markets" in the narrowly economic sense of the term.

SUMMARY

In this chapter we have examined two "experimental animals" which provided good testing grounds for structural analytic ideas and methods: community-elite networks and markets. The first of these problem-areas was created when structuralists succeeded in moving the debate over the social bases of power in communities away from a concern with individual power holders and toward a more systemic and sociologistic concern with the systems of linkages between and among organizations and groups. Laumann and his colleagues have developed a way of using smallest space models to examine properties of this structure which can provide detailed answers to many of the specific questions raised by both pluralist and antipluralist writers. This involves the creation of collective actors whose distances from one another can be plotted and used as an indication of consensus and disagreement among elite groups. Breiger was able to replicate the results of the "Altneustadt" study using a form of algebraic analysis called blockmodeling. This technique also yielded detailed mappings of the particular forms of consensus/disagreement at play among community leaders.

The second problem-area we dealt with, market structure, has not been fully explored as yet. Conventional notions of a market, however, are often too abstract, limited, and difficult to apply to "real world" situations. Structuralists have made some headway in defining markets in concrete terms. Present indications are that separate structuralist models of markets may be converging toward a general socio-economic theory.

EXERCISES

(Select one of the following)

1. Select one of the methods of studying community-elites described in this chapter. Conduct a small-scale study of your university "community." How would you choose your starting sample? Who would you choose as informants or which organizations would you select as "pivotal"? After you have collected your data (a) create a graph showing overlapping memberships between key organizations and (b) draw the dual of this graph. Analyze each graph separately, using one of the measures or methods described in this or earlier chapters. Describe your results and compare them to those obtained in one or more of the classic community-elite studies we examined here.

2. Focusing on the first or second dean's office problem, add in or take out some ties from the datagraphs. Create an image graph reflecting a set of hypotheses about this new structure. Block the data on the basis of this image. How closely does it conform to the blockings shown in Figure 4.7? If it is quite different, why? If not, why not?

3. Write a brief methodological critique of one or more of the early community-elite studies cited in this chapter. If possible—where actual data is given—reanalyze these data using one of the structuralist measures or techniques described in this volume. If it is not possible for you to do this, describe, in detail, the methods or techniques you would have employed and the results you would have expected to obtain.

4. Draw the graphs corresponding to Breiger's hypothesized image matrices (Figure 4.8). Which ties must be represented in a blockmodel? How different are Breiger's hypotheses? How "restrictive" are the models? In each case, indicate how a given model could be falsified. Is it consistent? If not, why not? Then create "your own" image graphs reflecting alternative hypotheses about community-elite structures. (You may find it useful to go to the library and look at other studies of community-elites to get ideas for these.)

ADDITIONAL SOURCES

Breiger, Ronald L.; Scott A. Boorman; and Phipps Arabie. "An Algorithm for Clustering Relational Data, with Applications to Social Network Analysis and Comparison with Multi-Dimensional Scaling." *Journal of Mathematical Psychology* 12 (1975): 328–83.

Breiger, Ronald L., and Philippa E. Pattison. "The Joint Role Structure of Two Communities' Elites." *Sociological Methods and Research* 7 (1978): 213–26.

Burt, Ronald S. "Positions in Multiple Network Systems, Part One: A General Conception of Stratification and Prestige in a System of Actors Cast as a Social Typology." *Social Forces* 56 (1977): 106–31.

Marsden, Peter V., and Edward O. Laumann. "Collective Action in a Community Elite: Exchange, Influence Resources, and Issue Resolution." In R. J. Liebert and A. W. Imershein (eds.), *Power, Paradigms, and Community Research*. Beverley Hills: Sage, 1977, pp. 199–250.

5

The Future of Structural Analysis: Some Conclusions

This book has briefly traced the development of structural analysis from a set of primitive intuitions about social process—and correspondingly rudimentary methods of data analysis—into a distinct, though incomplete, transdisciplinary research paradigm. Since my intention here was to provide an introduction to, rather than an exhaustive treatment of, the area, I have emphasized those particular theoretical developments, methodological advances, and exemplary applications of these which have set this new paradigm apart from other, sometimes closely related, approaches to social inquiry.

While useful heuristically, this line of attack has obviously entailed hard compromises. Each of the antecedent traditions which contributed to the emergence of structural analysis, and, indeed, all the work within the area itself, could not be treated with equal depth and rigor. Thus, contributions which, while important in their own right, have been less central to the *evolution* of the paradigm, per se, have been dealt with in notes or excluded entirely.[1] In addition, more technical and more recent literature has not been stressed except where it is essential to a clear presentation of core ideas and key methodological developments.[2] Burt, Freeman, Leinhardt, Wellman, and others have provided recent review articles, collections, or annotated bibliographies which may be used as a guide by readers who wish to explore some of these issues more systematically.[3]

Rather than try to backtrack and fill in these gaps at this point, this chapter will address a different kind of question. Previous chapters have been preoccupied with the specific intellectual trends that led to the emergence of structural analysis in the 1950s, 1960s, and 1970s. Here we will shift toward a more general focus on the constituent elements, latent organization, and implicit research agenda of this new paradigm and, on the basis of this discussion, attempt to forecast its impact on the traditional disciplines and research areas which it crosscuts. First, we will draw together some of the implications of the framework that has emerged from structuralist research practice during the last several decades.[4] Next we will critically evaluate the substantive and methodological directions most strongly suggested by the internal dynamics and overarching structure of this paradigm. Finally, we will tentatively examine the relationships between structural analysis and allied approaches to social inquiry.

STRUCTURAL ANALYSIS: INTERPRETATIVE SCHOOL OR PARADIGM?

Kuhn argues that sciences typically pass through several stages before they come to be dominated by a single paradigm, which then sets the agenda for researchers, provides a standard for training new practitioners, and establishes interpretative schema for assessing the relative weight assigned to new contributions.[5]

Before this paradigmatic or *normal* stage is reached, various schools of thought contend. No single framework is so widely recognized that it precludes "overt disagreements over fundamentals."[6] While research may be guided by theories, methods, and bodies of fact which collectively function like a true paradigm, no set of results or "exemplary observations and experiments" is, in itself, so persuasive that it precludes alternative modes of scientific activity.[7]

Once a science has been "normalized," however, it becomes relatively "closed," and the scientific community formed around its central paradigm becomes resistant to fundamentally new ideas and interpretations.[8] This, in turn, allows members of that scientific community to focus on a range of clearly defined and anticipatable problems, which can then be solved in a relatively straightforward way.[9] Closure of this sort, Kuhn argues, is essential to scientific advance. At the same time, however, since normal science assesses new findings in terms of how well they fit with orthodox interpretations, anomalous results are often treated as "failed experimentation," or simply as "special cases" which do not require a major modification of the central paradigm.[10]

Only when these anomalies become overwhelming, or when a dominant paradigm is challenged by a set of events external to the scientific community involved, is it possible for a new paradigm to gain ascendency.[11] Scientific revolutions, as Kuhn uses the term, occur when this happens: when an old paradigm "fails" and a new, and presumably more appropriate, paradigm replaces it.[12]

Throughout this book, I have argued that structural analysis does not represent a simple extension of existing social science paradigms, but, instead, signals the beginning of a scientific revolution. This statement may seem premature because, in multiparadigm disciplines such as sociology, anthropology, and so forth, the appearance of a new approach does not immediately, or even rapidly, occasion the kind of replacement of old paradigms that Kuhn observes in the physical or natural sciences.[13] Since divergent and even conflicting frameworks exist, it is easy to confuse the emergence of a radically new paradigm with the appearance of yet another interpretative school.

My argument here, by contrast, has been that while structural analysis has clear roots in a number of existing traditions, it proposes so distinct a departure from their typical modes of theory construction, tools, and customary research practice that it is best thought of as a qualitatively different approach. This perception has been reinforced by a number of sociologically important developments. First, during the last several years, structural analysts have created a highly specialized language in which to communicate concisely and technically with one another. As a matter of fact, much of this book has been taken up with the task of introducing readers to this language and to the conceptual apparatus that underlies it. Kuhn maintains that as a framework or paradigm matures its practitioners eventually reach a point where they have no choice but to create a vehicle of this sort for effectively sharing ideas and results with other members of their *scientific community*.[14] While the mere existence of a technical language per se cannot be taken as evidence that a particular research group has created its own paradigm, it is a powerful indication that a coherent and well-organized research group has been formed around some set of ideas, and that at least one criterion has been developed to define its boundaries.

Second, structural analysts have been concerned with a range of issues (e.g., the boundaries within and between levels in systems, patterns of relationships between sets of like elements, transitivity or indirect effects between nodes and their centrality or strategic location) which are outside the purview of conventional social science. The modeling and measurement techniques structuralists have developed or refined during the last several years—algebraic modeling, multidimensional scaling, "small world" models, clique and cluster detection procedures, graph theoretic measures, and statistical distributions of small-scale structures—have, as a result, been quite different from those included in a traditional graduate program in the social sciences. Thus, until recently, practical training in structural analysis has been provided at only a few places, and these universities and their students have tended to dominate the field.[15] But as graduates have gone on to teach at other institutions, the circle of practitioners with more than a passing knowledge of the technology of structural analysis has increased rapidly, adding both breadth and depth to the opportunities for graduate education in the area. Since Mullins and others have stressed the importance of graduate training in establishing paradigms within disciplines,[16] we can take this as further evidence that structural analysis has emerged as at least one among several alternative frameworks.[17] Of course, the availability of a wide spectrum of opportunities for graduate training in structural analysis does not in itself constitute proof positive that it may be seen as a paradigm as opposed to merely an interpretative school.

The third and final set of sociologically important clues as to the present status of structural analysis has to do with its degree of *institutionalization* within the scholarly world. Typically, this is signaled by the establishment of publications, formally constituted associations, special meetings or conferences, and recognized "slots" or allocations of time on the programs of the annual meetings of major professional bodies. In each of these senses, structural analysis has "come of age": both a journal, *Social Networks*, and a journal/newsletter, *Connections*, began publication in the late 1970s; the International Network for Social Network Analysis (INSNA) was founded; conferences or meetings dealing with developments within the field have become commonplace; and social network research has become an established topic at meetings of national and international sociological associations.[18] Here again, however—even taken together with the other evidence we have—these facts do not conclusively demonstrate that structural analysis represents a new paradigm rather than simply a novel method of gathering and assessing data which can be absorbed within existing frameworks. Put another way, even if we grant that structural analysis constitutes a *socially* cohesive body of work, we must also demonstrate that its *intellectual* foundations are sufficiently sound and well developed so that it can "stand alone."

The case for the evolution of structural analysis from a branch or interest area within conventional social science into a fullfledged paradigm therefore must ultimately be based on its ability (a) to cope with imperfectly understood or "marginalized" phenomena, that is, those inadequately treated within traditional frameworks, and (b) to deal more convincingly or robustly with conventionally interpretable material. While incomplete, the evidence thus far is unambiguous: *all*, or very nearly all, conventional social science, while sometimes acknowledging the systemic or larger-than-individual nature of social reality, rests on the methodological premise that social events take place in isolation or relative isolation from one another. Hence, they may be sampled as though they were independent, counted, aggregated together to form distributions, and statistically assessed using models that preclude or obscure detailed examination of the degree to which elementary units interact with one another. By contrast, the core of the structural analytic approach, as we have seen, lies in its willingness to deal with the kinds of complex and interrelated phenomena that almost all interpretative schools and paradigms in the social sciences studiously avoid. It approaches these problems by constructing analogs to an intricate and complex "reality," which it explicitly models through isomorphic or homomorphic mappings. These are then analyzed sociologistically and in a unitary way, dealing directly with the interplay of roles and positions at each of several layers within complex structures. In effect, structural analysis aims at capturing the holistic nature of social structure in a theoretically sophisticated and technically advanced fashion. While a variety of conventional interpretative schools and paradigms use terms like *systems* or *structure* which convey the sense that they are concerned with the overarching set of patterned relationships between parts of a social structure or system, no other paradigm has effectively embodied this idea in a theoretically specific and methodologically rigorous framework. In this sense, at least, structural analysis represents something quite unique within the spectrum of approaches currently adopted by social scientists.

A number of studies, moreover, support the belief that structural analysis has also succeeded in reinterpreting conventionally understood phenomena in new and convincing ways. Briefly, standard demographic analyses utilize a variety of "choice/constraint" models in which biological facts or limitations are the principal source of the "structure" implied in the modeling process.[19] Using a combination of simulation and more conventional techniques, however, Howell was able to show that an extremely close approximation to the demography of a population of hunters and gatherers (the Dobe !Kung) may be generated given a knowledge of its marriage structure.[20] While Howell's work took place in an almost ideal arena for examining the interplay of the "shape" of systems and the behavior of actors, it is nonetheless a remarkable demonstration of how structural analytic methods can be used to discover the processual dimensions underlying aggregate phenomena.[21] Peter Carrington's model of the relationship between the density of interlocking directorates and corporate concentration, Boorman and Levitt's explanation of patterns of genetic evolution, and a number of other recent studies share a common focus on systemic questions of this kind.[22]

Thus, on balance, it is clear that structural analysts now constitute a coherent research group pursuing common issues, using a set of common concepts and tools, and converging around a set of observably related types of phenomena. Some of these have been largely ignored by conventional social science, and some have not been dealt with adequately. On this basis, at least a prima facie case may be made that structural analysis now constitutes a paradigm in the specific sense of a body of theory, methods, and results which provides "model problems and solutions to a body of practitioners."[23] However, as we have noted throughout this book, it is a paradigm which is not yet completely formed or fully articulated. While structuralists can now point to a considerable body of "exemplary results and observations," only a few of structuralism's central theoretical questions have been empirically explored in any depth. Moreover, the relationships between the constituent parts or elements of this paradigm itself have yet to be articulated in detail. While a number of books and articles have begun to address some aspects of these questions, structural analysis, per se, has not yet been adequately theoretically codified.[24]

MORPHOLOGY, BEHAVIOR, AND STRUCTURE IN SOCIAL SYSTEMS

Textbooks—or even quasi texts such as this—are not really the proper place to propose a major theoretical synthesis. As a rule, readers who are just being introduced to a field find detailed technical exposition tedious. Practicing scientists, by contrast, demand just this sort of treatment before they are willing to take a new approach seriously.

Throughout this book, however, I have tried to balance the interests of genuine novices against those of practitioners who are merely unfamiliar with structuralist concepts and techniques. Therefore, it seems appropriate at this point to inventory at least the core ideas and animating concerns out of which this larger whole is emerging. While the precise way in which various elements will be "brought together" into a structuralist synthesis is still obscure, its broad outlines are beginning to become clear.

During the last three decades, structural analysts have dealt directly with three aspects of social structures: their morphology, or general form; the interactions among their elements, or behavior; and the patterned or persistent effects of these two, which we refer to as structure. Each of these corresponds to an analytically distinguishable dimension or dimensions of social process. Each, therefore, mutually conditions the others.[25] A system's morphology, for instance, fixes the limits in terms of which its elements interact (behavior). The *behavioral repertoire* of elements establishes, in turn, their expected range of variation under specified conditions. Over time, these limits and this behavioral repertoire fall into a pattern which is persistent and mutually reinforcing: the structure of that system. It is relatively resistant to changes in its scope, functions, and role organization.[26]

Much early work in structural analysis was preoccupied with devising ways of capturing the notion of *system morphology* in a theoretically consistent and rigorous fashion. As we mentioned briefly in Chapters 1 and 2, morphology is the abstract form of a system, that is, its organic organization. Since scientists can only examine it in reduced and simplified form, describing it involves two steps: mapping observed features of "reality" into a model, and then uncovering the overarching or global pattern in the relationships among elements. Since mapping involves a translation of one domain into another, interpretation is always involved. Hence, each analyst must decide, in advance, which set of units will constitute the primary focus of his or her model. These become "nodes." "Relations" are then operationally defined so as to capture the *embedding dimension* of that structure: the empirically ascertainable or measurable activity which most closely reflects the social process or processes a given analyst wishes to study.[27]

Since there is no universal method of determining how this translation ought to be made, each researcher must be guided by an explicit theoretical rationale which justifies the particular mapping rules he or she has followed.[28] Once these decisions have been made, however, the complex patterns of relationships between the parts of a structure can be explored, as we have seen, in a more or less formal fashion. Thus, while there are no "shortcuts"—and every analysis presumes a set of animating theoretical ideas and working hypotheses—these can be directly represented in the modeling process itself.

The abstract representation or image of the morphology of a system which results from this procedure, then, is *not* arbitrary: it incorporates a set of theoretical assumptions and is conditioned by a body of empirical evidence, that is, the observed relationships between nodes. Consequently, two structures that have been modeled following the same conventions can be formally and rigorously compared and inferences drawn about them in a way which is analogous to our treatment of distributions of "independent" events, for example, live births or incomes.[29]

The purpose of these comparisons is to learn something about the *constraints* which the form of a system imposes on the behavior of its component parts. Structuralists argue that morphologically similar systems constrain the behavior of their elements in formally similar ways: "common form bespeaks common constraint." As social structures evolve, their elementary units are formed in dynamic interaction with one another. Corporations, for instance, adapt to the market behavior of other corporations. In practical terms, this means that both the characteristic behavior (behavioral repertoire) and relationships among the elements represented in a structural model are the products of a set of *simultaneous processes* in which these behaviors and relationships have been involved in the past. The *patterning* of these relationships, as a result, may be treated as a complex index of the *mutually determined* actions and *jointly recognized* systemic constraints under which a given system developed.

Ideally, assuming that we can adequately model the constraints implied in the morphology of a structure and describe the behavioral repertoire of its elements, we ought to be able to anticipate the typical "behavior under constraint" of the elements. In fact, a well-developed branch of mathematical social science called *game theory* has been highly successful in deriving a set of models to explain the behavior of actors *given* a knowledge of the "rules of the game" (contraints), the utilities they attach to various outcomes, and the benefits likely to accrue to them ("payoffs") from following different courses of action ("strategies").[30]

While relatively few researchers have explicitly made this connection between structural morphology and game theoretic models of behavior under constraint, there is an obvious affinity between the two. As we noted earlier, in cases where structuralists have recognized this—as, for instance, in Boorman's formalization of Granovetter's job-search model or White's reassessment of classical market theory—they have been highly successful. It is safe to assume that further extensions of work of this kind will take place and that, because of the part it will play in the development of structural analysis as a whole, this work is likely to constitute the single most important growing edge of work in the area.

At this point, however, no structural analyst can claim to have produced a model which will satisfy the normal canons of scientific research equally well in terms of each of these three dimensions of social process. Moreover, few have even attempted to deal with all of them. Despite this, as we have seen in earlier chapters, examples of models which treat at least two are well represented. Laumann et al., for instance, have successfully associated the coalitional structure of elites (morphology) with the influence they seem to have in local affairs (structure). In the course of doing this, they were able to relate both of these dimensions to at least a description of patterns of elite interaction (behavior).[31] Breiger's reformalization of this behavioral level in terms of a blockmodel would, hypothetically at least, provide a good beginning point for the sort of generalized strategic model that Boorman developed to explain Granovetter's job-market data.[32] Similarly, Wellman's typology of hypotheses about the nature of community morphology could be reinterpreted in behavioral terms and compared to data derived from a variety of structural contexts. Burt's formalization of the relationship between the constraints under which firms operate and their links to different sectors of an economy provides another example of an attempt to link up models of these different processual dimensions.[33]

Thus, while it would be premature to predict precisely how structuralists are likely to proceed in the future, they are sensitive to the problems or challenges posed by each of these dimensions for the others. While one can point to examples of even very early structuralist research where the relationships between group-form and behavior were taken as problematic—as, for instance, in Coleman's seminal study of adolescent peer groups[34] or Charles Tilly's research on collective violence[35]—much recent effort has been spent creating more adequate and precise descriptions of morphology, behavior, or structure in relative isolation from one another. The latest trends, however, appear to reflect a movement back toward a focus on these initial and, in some respects, fundamental interdimensional or interprocessual problems.[36]

Given the developments which have occurred in the intervening years, this new wave of research is likely to be highly formal. Douglas White, for instance, has developed a modeling technique—called "material entailment" or, simply, "entailment"—which may be used to establish the extent to which theoretically possible transitivity between nodes can be demonstrated empirically. Since his technique can deal with and, in fact, demands multiple measures of connectivity, it holds out the possibility of a general solution to the methodological problem posed by Berkowitz et al., namely, the extent to which continuous or nearly continuous networks can be partitioned into theoretically meaningful subunits.[37] If "entailment" can be extended along these lines, and if structural analysts can develop appropriate interpretations of the clusters of like nodes which it is designed to uncover, it could also open the door to a related and theoretically critical problem which has haunted systems theory since the 1940s and 1950s: the specification of *functional,* as opposed to natural or intuitive, subdivisions within larger systems. Given that "entailment" itself has a direct algebraic interpretation, there is a strong possibility that some headway will be made in the formalization of this more general question in the next several years.

Another formal approach to the explanation of the joint effects of more than one social process, simulation, is likely to yield interesting solutions to otherwise intractable problems. While Howell and Lehotay's work clearly demonstrates that complex structural-demographic problems can be examined using this technique, there are ample indications that it can be applied to a variety of other structural analytic questions as well. John Delaney's equally impressive simulation of the emergence of social structure from personal contact networks, for instance, shows that computational solutions can be used to clarify the consequences of extremely complex alternative structural models.[38] Since a number of structural analysts are now modeling systems where patterns of interaction among elements and end-states can be specified rather precisely, but where oscillations toward these states are obscured by the sheer number of paths they may follow, it seems plausible to predict that there will be an increasing interest in computer simulations of these types of structures, as well.

In the medium term, new insights into the ways in which morphology and behavior condition one another are most likely to grow out of extensions of the algebraic modeling techniques developed by White, Boorman, Breiger, Heil, and others.[39] Carrington, Heil, and Berkowitz have created a continuous measure of the goodness-of-fit of image graphs to datagraphs in blockmodeling. While their measure, *b,* has not yet been associated with a probability distribution—and, hence, cannot be treated as though it were, i.e., used in a statistical test—it does allow researchers to compare alternative blockings to one another.[40] Since, in the perspective supported here, the principal purpose of science is to "uncover" relationships among sets of events, a measure such as this is an essential component of any line of scientific inquiry. Without it, researchers are forced simply to propose models, without a means of establishing their relative ability to summarize obtained data. While, ideally, we would like to be able to rigorously assess the likelihood that given models differ significantly in their goodness-of-fit to data—from one another and from chance (the underlying purpose of statistical tests)—Carrington, Heil, and Berkowitz's work is an important first step toward this.[41] Thus, one obvious direction structural analysis may take in the future is toward developing statistical tests based on b or other b-like measures.[42]

Another algebraic approach, developed by R. H. Atkin, draws a distinction between what is referred to as *backcloth*—structural relations (morphology)—and *traffic*—the channel-bound flow of activity (or behavior)—within a system.[43] The purpose of his technique, called q-analysis, is to relate one to the other in order to discover which "traffic" patterns are permitted by the "backcloth" and how various structural characteristics, such as q-holes, *generate* these flows. In principle, of course, this is precisely the direction in which I have argued structural analysis ought to be moving. But some of the implications of q-analysis—which derives from algebraic topology—are not well understood at the moment, and it is still unclear whether or not simpler graph theoretic formulations of this problem could not serve the same purpose.[44] Thus, while Atkin's q-analysis promises to be an important extension of existing structural analytic methods, further research is needed before we can be sure that it is the most parsimonious way to explain the interface between morphology and some forms of behavior.

FINALE: TOWARD A UNIFIED SCIENCE OF SOCIETY

When general systems theory emerged in the mid-1950s, some of its proponents were hopeful that artificial barriers between disciplines would begin to crumble and that a new, broadly based and synthetic approach to science would take their place.[45] In retrospect, while this has clearly not happened, no scientific field can claim to be uninfluenced by its core concepts and paradigms. At least the terminology or *language* of general systems theory has been incorporated into all, or very nearly all, areas in the physical, natural, and social sciences. Its key notions, such as "feedback" or "goal-oriented systems," have even become part of the popular culture.

General systems theory itself, however, remains a *collection* of ideas rather than a systematic body of work, largely because of its lack of focus on a set of well-defined empirical problems.[46] In some sense, structural analysis may be thought of as the most direct and concrete manifestation of the core ideas of general systems theory within the social sciences: *holism*, or the importance of explaining phenomena not only in terms of their components, but the entire set of relations between them; *scientific integrality*, or a conceptualization of science as a single form of inquiry; and *natural unitarism*, or the assumption that analogous patterns of order may be found in different domains within "nature." But because of its clear focus around a relatively restricted set of "experimental animals" and theoretical issues, structural analysis has managed to avoid the vagueness or lack of precision which is characteristic of much contemporary general systems research.[47] Thus, while structural analysis bears an obvious familial relationship to systems theory, it is clearly not a mere branch or subdivision within it.

Similarly, while structural analysis frequently employs mathematical reasoning or tools, it is clearly tied to a set of theoretical traditions, principally in sociology and economics. By contrast, mathematics itself—or even mathematical sociology—is, for all intents and purposes, theoretically neutral: one can express *any* set of logically consistent theoretical statements in mathematical terms. Thus, while structural analysis shares an obvious affinity with mathematical sociology, they are not the same.

Perhaps structural analysis's closest cousins in the scholarly world are the other forms of "structuralism" which have grown up in linguistics, anthropology, and psychology. Readers who are familiar with the writings of Ferdinand de Saussure, Roman Jakobson, Morris Halle, and Edward Sapir in linguistics, anthropologists such as Franz Boas and Claude Lévi-Strauss, or Jean Piaget's work in psychology will immediately be struck by similarities between the core ideas and organizing concepts they employ and those presented here.[48]

These apparent "correspondences," to borrow Lévi-Strauss's term, are not accidental: *all* contemporary approaches to scientific inquiry which refer to themselves as "structural," "structuralist," or "structural analytic" share an abiding concern with patterning in events, with relations among the elementary components of systems or structures, and with the primacy of systemic transformations in shaping processes. Each form of structuralism has, moreover, obviously borrowed both concepts and language from the others.[49] But here the similarities between the other forms of structuralism and structural analysis end. Critically, structural linguistics, structural anthropology and structural psychology confuse what philosophers refer to as *epistemology*—or the branch of knowledge which has to do with how we come to know things—with *ontology*—the study of "being" or "existence." Epistemologically, all structuralists agree that "structure" is a construct—that is to say, it is created by people to explain or model their observed world. But, singly and collectively, the other forms of structuralism—particularly in the social sciences—argue that *there is no reality* apart from what human consciousness itself can construct, and that "structures" are therefore closed, self-regulating systems which the human mind imposes on a pliable substratum of observations. Consequently, *only* relations between objects are important, since objects have *no* meaning apart from their context.[50]

Throughout this book, I have emphasized what structural analysts refer to as the "dual" nature of social processes: that there is a dynamic interplay between the local and particular manifestations of social processes and their global and general aspects. Epistemologically, or, in a more restricted sense, methodologically, this has been evidenced by the development of techniques, such as algebraic modeling, which require that both "positional" and "relational" phenomena be explained in the same terms and at the same time. Ontologically, this has been expressed in a notion of science that distinguishes between models and the "reality" they seek to capture or represent. Structural analysts would, therefore, argue that nonstructural analytic "structuralism," because it sees relational properties of systems as uniquely determining the characteristics or attributes of their members, has disregarded *half* of what is essential to understand about social life. Put another way, structural linguistics, à la Saussure and Jakobson, structural anthropology as interpreted by Lévi-Strauss, and Piaget's psychology ultimately rest on ideas or consciousness (they are, to use the philosophic term, *idealistic*), whereas structural analysis requires that scientists ultimately come to terms with an observed material reality. While it is important to recognize that scientists model this reality to accord with particular purposes they have in mind, nonetheless, they are constrained by the necessity to observe and record data which validates their constructs. Only if one argues—as some nonstructural analytic structuralists do—that all observations and perceptions are simple by-products of innate internal organization of mind is it possible to escape this.[51] Structural analysis tries to deal with, rather than avoid, the issue.

These fundamental differences between general systems theory, mathematical sociology, generally, and idealist structuralism, on the one hand, and structural analysis, on the other, reinforce our assessment of its unique place within contemporary social science. During the last several years, its impact on the fields and disciplines it crosscuts has been greatest in those specific problem areas out of which it has drawn its "experimental animals." But it has not been exclusively confined to them: there is evidence of the gradual assimilation of specially structural analytic ideas and techniques in biology, management studies, history, clinical and social psychology, and linguistics, as well as more traditional areas of concentration, such as sociology and economics. Moreover, since structural analysis is not a field in the conventional sense of the term, this kind of proliferation across disciplinary boundaries and into the hitherto sacrosanct preserves of specialized research groups is likely to continue.

Two important dynamics are evident in this dispersion or propagation of structuralist ideas, models, and techniques thus far. First, to date it has tended to have a differential impact on research areas where the organization of, or relationships between, units is problematic. Thus, for instance, structuralists have been dealing with problems in "industrial organization" for quite some time, but have only recently begun to tackle problems in "microeconomics." Small groups, bureaucratic organizations, scientific communities, and friendship circles have been elaborately researched, but psychiatric groups, therapeutic settings, and prison populations have not. While some structuralists have been dealing, and may be expected to deal, with these sorts of milieux in the future, the weight of current research is not moving in this direction.

Second, as structuralists have entered new areas, the tendency has been to absorb or assimilate ideas, assumptions, and models from these sites and disciplines into their own work. Thus, the kind of starkness or, more favorably, detachment which was evident in many of the earliest formal structural analytic models has given way to a closer interest in the research issues and formal techniques within these fields and specialties.

If both of these trends continue—the dispersion of structuralist approaches into research areas where "organization" is problematic, and the infusion of a range of ideas drawn from a variety of fields into structural analysis proper—then this new, transdisciplinary paradigm may well lead to the formation of the kind of *general* social science paradigm which some of its antecedent traditions anticipated over two decades ago. If it does, we may yet see the consolidation of social scientific inquiry around the kind of authoritative paradigm which Kuhn observed in the physical and natural sciences.

EXERCISES

1. Begin with the sources in the notes to one section of this book. Go to the library and find the books and articles cited in these sources. Create a mapping or a series of mappings of the citation network among them. (HINT: a "source" includes both an author's name and the name of an article.) Cluster these sources together on the basis of some criterion and create supernodes (e.g., include within one

supernode all sources that have at least one author in common). Then, taking each connected component of the new network of ties among supernodes separately, calculate the density of ties using the standard density measure.

$$\frac{2a}{N(N-1)}$$

Redraw the boundaries around sets of nodes in such a way as to improve this density, if possible. On the basis of this, can you draw conclusions about which authors and/or sources form the "core" of structural analysis and which are at the "periphery"? (This assumes, of course, that the initial citations were to work in structural analysis in the first place.) Could you improve the accuracy of this calculation by taking multiplexity into account?

2.   Using the data gathered in 1, above, distinguish between positive, negative, and neutral citations. Does this improve the precision of your modeling? If so, how? If not, why not? Employing one or more of the techniques described in this book, create a model which uses these distinctions between the valence of citations. Explain what you have done.

3.   This chapter distinguishes among three dimensions of social process: morphology, behavior, and structure. Create *simulated* datasets for each of these three aspects of a particular structure and develop a verbal or formal model explaining their relationships to one another.

ADDITIONAL SOURCES

Abell, P. "Measurement in Sociology: I. Measurement and Systems." *Sociology* 2 (1968):1–80.

Abell, P. "Measurement in Sociology: II. Measurement Structure and Sociological Theory." *Sociology* 3 (1969): 397–411.

Abell, P. *Model Building in Sociology*. London: Weidenfeld and Nicolson, 1971.

Arrow, K. J.; S. Karlin; and P. Suppes (eds.). *Mathematical Models in the Social Sciences*. Stanford: Stanford University Press, 1960.

Boorman, Scott A., and Phipps Arabie. "Algebraic Approaches to the Comparison of Concrete Social Structures Represented as Networks: Reply to Bonacich." *American Journal of Sociology*, forthcoming.

Coleman, James S. *An Introduction to Mathematical Sociology*. New York: Free Press, 1964.

Fararo, Thomas J. *Mathematical Sociology*. New York: Wiley, 1973.

Gray, W., and N. Rizzo (eds.). *Unity through Diversity: Festschriff in Honor of Ludwig von Bertalanffy*. New York: Gordon & Breach, 1971.

Laslo, E. *Introduction to Systems Philosophy*. New York: Gordon & Breach, 1971.

Wiener, Norbert. *Cybernetics*. New York: Wiley, 1948.

Notes

1: What Is Structural Analysis?

1. For an excellent, though brief, review of the history and uses of the concept of "structure" in various fields, see Terence Hawkes, *Structuralism and Semiotics* (Berkeley and Los Angeles, Calif.: University of California Press, 1977). Jacques Ehrmann (*Structuralism* [Garden City, New York: Doubleday, 1970]), Richard T. and Fernande M. De George (*The Structuralists: From Marx to Levi-Strauss* [Garden City, New York: Doubleday, 1972]) and Michael Lane (*Structuralism: A Reader* [London: Cape, 1970]) have assembled intelligent collections of the key literature on structuralism, per se. Hawkes, and others trace the use of the concept in social sciences to the Italian writer Giambattista Vico, whose central work, *Principi Di Scienza Nuova*, first appeared in 1725. See Thomas Goddard Bergin and Max Harold Fisch (eds.), *The New Science of Giambattista Vico* (Ithaca: Cornell University Press, 1970).

   The fifth chapter of this book contains an outline of forms of structuralism as they are interpreted today in linguistics, anthropology, and psychology. For present purposes, we will beg the question of the relationship between these forms of structuralism and structural analysis. Unless otherwise specified, the terms *structural analyst* and *structuralist*, and *structural analysis* and *structuralism* will be used synonymously.

2. Note the similarity between this definition of *structure* and Edward O. Laumann and Franz Pappi's. "A *social structure* [is] defined as a persisting pattern of social relationships among social positions" (Edward O. Laumann and Franz Pappi, *Networks of Collective Action: A Perspective on Community Influence Systems* [New York: Academic Press, 1976], p. 6). Laumann and Pappi, correctly, draw a distinction between those complex sets of relationships we refer to as social *systems* and the more limited constructs we derive from them, called *structures*. This distinction is maintained here. Structure, per se, can be a *property* of either or both of them (cf. Leon Mayhew, *Society, Institutions, and Activity* [Glenview, Ill.: Scott, Foresman, 1971]; Peter M. Blau, "Parameters of Social Structure," *American Sociological Review* 39 (1974): 615–35).

   The notion of "paradigms," which will play an important role in much of what follows, refers to codifications of social theory and research practice which, to use Kuhn's formulation, "provide model problems and solutions to a body of practitioners" (Thomas Kuhn, *The Structure of Scientific Revolutions* [Chicago: University of Chicago Press, 1970, p. viii]). As such, paradigms consist of basic assumptions about the nature of "reality," as well as how we go about apprehending or interpreting it. Merton uses the term in a more restricted way, i.e., as referring to a set of broadly interrelated theoretical propositions and findings (Robert K. Merton, *Social Theory and Social Structure* [New York: Free Press, 1968], pp. 69–72). The chief difference between these usages is that Kuhn's definition of a paradigm incorporates notions of *ontology*, or the existence of things, as well as epistemology or methodology.

3. For a summary of the issues involved in "measurement" and its role in science, see Abraham Kaplan, *The Conduct of Inquiry: Methodology for Behavioral Science* (Scranton, Pa.: Chandler, 1964), pp. 171–214. Kaplan (op. cit., 1964, pp. 39–42) cautions against an over-literal acceptance of the assumptions of "operationism."

   In its most general or global sense, the term *structure* is sometimes understood to refer to any principle of ordering or organization discernible in a system. It is this sense of the term, for instance, which is used by economists when they refer to the "structure-conduct-performance paradigm" in economics (Joe S. Bain, *Industrial Organization* [New York: John Wiley & Sons, 1968], pp. 462–68). As I will argue throughout this book, this use of the term to refer to attributes of the elementary units within systems is exotic and misleading.

4. The mathematical basis of graph theory and network analysis is summarized well in Robert G. Busacker and Thomas L. Saaty, *Finite Graphs and Networks: An Introduction with Applications* (New York: McGraw-Hill, 1965), and, in more specific contexts, in Lee M. Maxwell and Myril B. Reed, *The Theory of Graphs: A Basis for Network Theory* (New York: Pergamon, 1971), and Frank Harary, Robert Z. Norman, and Dorwin Cartwright, *Structural Models: An Introduction to the Theory of Directed Graphs* (New York: John Wiley & Sons, 1965).

   Mathematicians usually use the term *networks* to refer to one highly specific class of graphs. Here we will follow the generally accepted *sociological* convention of defining a *network* as a set, S, of objects and a set, E, of weighted or unweighted edges on E. Since there are no restrictions on how these objects may be connected to one another, we set aside, for the moment, questions as to the precise type of edges involved, e.g., whether or not two can parallel one another. More mathematically precise terms will be introduced as needed.

5. The terms *system, element,* etc., which are used here, derive from the field or area referred to as *general systems theory.* A system consists of a set of strongly interacting "elements" which, taken together, form a complex, organized whole (Anatol Rapoport, "Mathematical Aspects of General Systems Theory," *General Systems* 11 (1966): 3–11; Ludwig von Bertalanffy, *General System Theory* [New York: George Braziller, 1968], p. 19). Direct application of these concepts to sociological problems has been made by Buckley (Walter Buckley, *Sociology and Modern Systems Theory* [Englewood Cliffs, N.J.: Prentice-Hall, 1967]). Less rigorous use of these and related terms, however, abounds in the social sciences. Here we intend their more formal and restricted meaning.

6. See J. Clyde Mitchell, "The Concept and Use of Social Networks," an excellent summary of the strengths and weaknesses of the earliest work in the area (in J. Clyde Mitchell (ed.), *Social Networks in Urban Situations: Analyses of Personal Relationships in Central African Towns* [Manchester: Manchester University Press, 1969], pp. 1–50). Barry Wellman provides an excellent critical discussion of both the early trends in, and subsequent development of, the field (Barry Wellman, "Network Analysis: From Metaphor and Method to Theory and Substance," *Sociological Theory* 1, forthcoming).

   My use of the terms *morphology, behavior,* and *structure* here follows from Mitchell (Mitchell, op. cit., 1969). Chapter 5 will deal with the detailed meaning and applications of these concepts.

7. A good synthesis of these "classical" approaches to the study of infectious disease dispersion is provided in Norman T. J. Bailey, *The Mathematical Theory of Infectious Diseases and Its Applications* (New York: Hafner, 1975).

8. See Anatol Rapoport, "A Probabilistic Approach to Networks," *Social Networks* 2 (1979/80): 1–18, for a concise discussion of these issues.

9. Rapoport, op. cit., 1979/80, p. 5.

10. Anatol Rapoport, "Nets with Distance Bias," *Bulletin of Mathematical Biophysics* 13 (1951): 85–91.

11. Anatol Rapoport, "Ignition Phenomena in Random Nets," *Bulletin of Mathematical Biophysics* 14 (1952): 35–44. Similar problems were examined where other kinds of "information" were involved. See Anatol Rapoport, "The Diffusion Problem in Mass Behavior," *General Systems* 1 (1956): 48–55; James S. Coleman, Elihu Katz, and Herbert Menzel, *Medical Innovation: A Diffusion Study* (Indianapolis: Bobbs-Merrill, 1966). Everett M. Rogers, *Diffusion of Innovations* (New York: Free Press, 1962) summarizes much of this early literature on the transmission and acceptance of innovations. Coleman, Katz, and Menzel's work was significant in that it systematically explored both the structure of information transmission and the structural preconditions for the adoption of new ideas or techniques. Also see Herbert Menzel and Elihu Katz, "Social Relations and Innovation in the Medical Profession: The Epidemiology of a New Drug," *Public Opinion Quarterly* 19 (1955): 337–52.

12. *Bias* is the extent to which a network (more properly, in this case, a *net*) departs from a random or baseline model. Parameters are introduced to explain these discrepancies. The theory behind the interpretation of bias in social networks was first put forward in Anatol Rapoport, "Nets with Reciprocity Bias," *Bulletin of Mathematical Biophysics* 20 (1958): 191–201; Anatol Rapoport and William J. Horvath, "A Study of a Large Sociogram," *Behavioral Science* 6 (1961): 279–91; and Caxton C. Foster, Anatol Rapoport, and Carol J. Orwant, "A Study of a Large Sociogram, II: Elimination of Free Parameters," *Behavioral Science* 8 (1963): 56–65. This work is synthesized and extended in Thomas J. Fararo and Morris H. Sunshine, *A Study of a Biased Friendship Net* (Syracuse: Syracuse University Press, 1964).

13. Wellman, op. cit., forthcoming.

14. Kaplan, op. cit., 1964, pp. 285–87, 306–10.

15. This aspect of structuralist thinking comes through most clearly in algebraic modeling. In the Lorrain-White method or blockmodeling, as we shall see, systems of relations are simplified ("reduced") and the role of elements deduced from this reduction. These methods will be treated extensively in Chapters 3 and 4. For a discussion of the theoretical issues involved, see S. D. Berkowitz and Greg Heil, "Dualities in Methods of Social Network Research" (Toronto: Structural Analysis Programme, 1980).

16. A *heuristic* is a device whose purpose is to stimulate investigation. The exercises in this book, for instance, are designed to play a heuristic role: to stimulate students to discover things about social networks on their own.

17. The term *multidimensional scaling* or MDS refers to a variety of techniques used in the social sciences to uncover patterns in data and render them in the form of a spatial representation. The mechanics employed in MDS will

be discussed in detail later in this book. For the moment, multidimensional scaling allows investigators to present important features of a set of data in terms of distances between points. Since the theoretical upper limit to the number of dimensions needed to do this perfectly is $N-1$, where N is the number of points, MDS techniques reduce the number of dimensions to the point where human beings can easily perceive the features of the data involved—usually two or three dimensions. R. N. Shepard, A. Kimball Romney, and Sara Beth Nerlove (eds.), *Multidimensional Scaling: Theory and Applications in the Behavioral Sciences* (New York: Seminar Press, 1972), presents both a history and a detailed overview of developments in the area.

18. John Paul Boyd, "The Algebra of Kinship" (doctoral dissertation, University of Michigan, Ann Arbor, 1966).

19. Harrison C. White, *An Anatomy of Kinship* (Englewood Cliffs, N.J.: Prentice-Hall, 1963); see Claude Lévi-Strauss, *The Elementary Structures of Kinship* (Boston: Beacon Press, 1969), esp. pp. 221–29.

20. John Paul Boyd, "The Algebra of Group Kinship," *Journal of Mathematical Psychology* 6 (1969): 139–67, presents this case most cogently.

21. See, in particular, John Paul Boyd and William Livant, "Some Properties and Implications of Lexical Trees" (Ann Arbor: University of Michigan MS, 1964); C. Flament, *Théorie des Graphes et Structure Sociale* (Paris: Mouton, 1965); Noam Chomsky, "Formal Properties of Grammars," in R. D. Luce, R. R. Bush and E. Galanter (eds.), *Handbook of Mathematical Psychology*, vol. 2 (New York: Wiley, 1963); Dorwin Cartwright and Frank Harary, "Structural Balance: A Generalization of Heider's Theory," *Psychological Review* 63 (1956): 277–93.

22. François Lorrain and Harrison C. White, "Structural Equivalence of Individuals in Social Networks," *Journal of Mathematical Sociology* 1 (1971): 49–80. Kim and Roush define a *semigroup* as a "set S provided with a function x*y from S × S to S, satisfying the associative law (x*y) *z = x*(y*z)." Thus, it is a simple algebraic system consisting of a set and a binary operation on that set, which is associative (Ki Hang Kim and Fred William Roush, *Mathematics for Social Scientists* [New York: Elsevier, 1980], p. 10). "Compound" relations are formed when separate relations, e.g., *A* and *B*, are combined into a single relation *AB*. The operation involved in doing this is referred to as *composition* (Lorrain and White, op. cit., 1971, p. 71).

23. According to Lorrain and White, two nodes are structurally equivalent if each relates to all other nodes in *precisely the same way* as the other (Lorrain and White, op. cit., 1971, p. 81). This definition excludes cases where each relates to *equivalent* others in the same way. For an expansion of the concept to include notions such as this, see Douglas White, "Structural Equivalences in Social Networks: Concepts and Measurement of Role Structures" (Irvine, Calif: School of Social Sciences, University of California, Irvine, MS, 1980) (prepared for the Laguna Beach Conference on Research Methods in Social Network Analysis).

24. The importance of indirect ties in regulating the behavior of elements of a social system will be discussed at length in Chapter 2. In algebraic modeling, indirect associations between nodes allow us to assess the degree to which localized properties may be generalized to the network as a whole. For a detailed discussion of the mathematical implications of this, see Lorrain and

White, op. cit., 1971, and François Lorrain, *Réseaux Sociaux et Classifications Sociales* (Paris: Hermann, 1975).

25. R. N. Shepard, "The Analysis of Proximities: Multidimensional Scaling with an Unknown Distance Function, I," *Psychometrika* 27 (1962): 125–40; R. N. Shepard, "The Analysis of Proximities: Multidimensional Scaling with an Unknown Distance Function, II," *Psychometrika* 27 (1962): 219–46; R. N. Shepard, "Metric Structures in Ordinal Data," *Journal of Mathematical Psychology* 3 (1966): 287–315; R. N. Shepard and J. D. Carroll, "Parametric Representation of Nonlinear Data Structures," in P. R. Krishnaiah (ed.), *International Symposium on Multivariate Analysis* (Dayton, Ohio, 1965) (New York: Academic Press, 1966), pp. 561–92; Clyde H. Coombs, *A Theory of Data* (New York: Wiley, 1964); Louis Guttman, "A General Nonmetric Technique for Finding the Smallest Coordinate Space for a Configuration of Points," *Psychometrika* 33 (1968): 469–506; Edward O. Laumann, *Prestige and Association in an Urban Community* (Indianapolis: Bobbs-Merrill, 1966); Edward O. Laumann and Louis Guttman, "The Relative Associational Contiguity of Occupations in an Urban Setting," *American Sociological Review* 31 (1966): 169–78; Joel H. Levine, "The Sphere of Influence," *American Sociological Review* 37 (1972): 14–27.

26. Laumann and Pappi, op. cit., 1976; Edward O. Laumann, Peter V. Marsden, and Joseph Galaskiewicz, "Community-Elite Influence Structures: Extension of a Network Approach," *American Journal of Sociology* 83 (1977): 594–631; Edward O. Laumann and Franz Pappi, "New Directions in the Study of Community Elites," *American Sociological Review* 38 (1973): 212–30; Edward O. Laumann and Peter V. Marsden, "The Analysis of Oppositional Structures in Political Elites: Identifying Collective Actors," *American Sociological Review* 44 (1979): 713–32.

27. See, in particular, William Carroll, John Fox, and Michael Ornstein, "The Network of Directorate Interlocks among the Largest Canadian Firms" (Downsview, Ontario: Institute for Behavioral Research, York University, 1977).

28. The question as to whether or not structural analysis constitutes a paradigm, in the formal sense of the term, is set aside for the moment and raised again in the last chapter.

29. One curious example of this was the relative lack of contact between researchers at Columbia—centering around Charles Kadushin—and other groups until the early 1970s. This occurred despite Kadushin's clearly relevant contributions to the literature in the mid-1960s ("The Friends and Supporters of Psychotherapy: On Social Circles in Urban Life," *American Sociological Review* 31 (1966): 786–802; "Power, Influence and Social Circles: A New Methodology for Studying Opinion Makers," *American Sociological Review* 33 (1968): 685–99). A partial explanation of why this occurred is that Columbia was so clearly dominated by other strong paradigms that it was not "the place to go" if one was interested in studying structural analysis, i.e, it was outside the referral network for students entering graduate school in the area. This is no longer true.

30. Nicholas Mullins (*Theory and Theory Groups in Contemporary American Sociology* [New York: Harper & Row, 1973], p. 254) presents a table showing the "place" of graduate training and "selected job locations" for a set of

"important structuralists." Both Harvard and Michigan appear frequently. There are data missing from the table, however, which would make this point even more clearly. Thus, for instance, Michigan should have appeared as Boyd's home institution and his place of employment in 1965. S. Berkowitz and H. Friedmann (Berkowitz) were undergraduates at Michigan. Other people who should have been included in the list were then teaching there. Similarly, if 1973 had been chosen as a data year—as opposed to 1972—then the pattern which emerged in the late 1970s would have come through more clearly. By that date, for instance, Berkowitz, Friedmann, and Howell, as well as Howard and Wellman, were at Toronto. Rapoport had moved there in the late 1960s or early 1970s. Granovetter moved back to Harvard and then to Stony Brook, where he joined Schwartz. Chase eventually arrived there. Other structuralists, at about this time, shifted to the University of California at Irvine and Santa Barbara. Freeman moved to Irvine in the late 1970s. Mullins was joined at Indiana by at least one ex-Michigander, David Knoke.

31. Notably at Cardiff, Edinburgh, Leicester and Oxford in the U.K., Munich and Cologne in Germany, Vienna in Austria, and the University of Amsterdam in the Netherlands.

32. I argue, following Kuhn, that all sciences are primarily defined by the way in which they frame problems, rather than by *particular* interpretations of their tasks. Merton, as we noted in note 2, adopts a more restricted definition of a paradigm and, hence, of a science. Thus, in my framework, a set of researchers who frame problems in a similar way can constitute a coherent "theory group" even if they diverge substantially in other respects. These issues will be dealt with in detail in Chapter 5.

33. See Anatol Rapoport, "The Uses of Mathematical Isomorphism in General Systems Theory," in George J. Klir (ed.), *Trends in General Systems Theory* (New York: Wiley, 1972). An *analog* is something which resembles or is like something else in terms of two or more attributes, circumstances or effects. The term *real world* is emphasized here because scientists have no way of apprehending events except through the construction of analogs. Thus, in effect, we are postulating the existence of a "real world" which we cannot verify directly. Things are said to be *isomorphic* if there is a one-to-one correspondence between them. Hence, for every part or relation in one, there is a corresponding part or relation in the other. Things are said to be *homomorphic* if many such things can be *reduced* to one, i.e., it is a many-to-one mapping.

34. Mitchell, op. cit., 1969, and von Bertalanffy, op. cit., 1968, stress the importance of "forms of correspondence" between systems. Here we have extended this idea to include purely formal aspects of the organization of systems. This is consistent with Mitchell's (op. cit., 1969) use of the term *morphology*.

35. See Kaplan, op. cit., 1964.

36. Note that there is no implication here that scientific reasoning is, in any sense, superior or inferior to any other form; it is merely asserted that they are different. Some important propositions about the world simply cannot be formulated in scientific terms, e.g., the present debate between Creationism and Evolutionism. This means, of course, that science is not, contrary to popular belief, an all-encompassing worldview.

37. The term *psychologistic* should not be confused with *psychological*. As we use the term here, *psychologistic* simply implies that the weight of a given set of inferences proceeds from parts to wholes. If psychologists reason in the opposite way, their direction of inference is *not* psychologistic, e.g., some social psychologists. Most psychologists, however, do draw inferences about wholes from parts—hence the term.

38. See Berkowitz and Heil, op. cit., 1980, for a more complete discussion of this issue. We will examine it again when we discuss clustering procedures.

39. "Higher order structures" may be of two kinds: (1) sets of elements which function, for all intents and purposes, as if they were one unit, i.e., as "supernodes" and (2) sets of elements which are combined in different ways at higher orders of generalization, e.g., ones mapped together on the basis of structural equivalence. This dichotomy will become important in Chapter 4 when we compare clustering and blocking as operations.

40. See von Bertalanffy, op. cit., 1968, pp. 3–29, for a discussion of the foundations of general systems theory. The preface to the enlarged edition cited above draws explicit parallels between developments in general systems theory and some forms of structuralism.

41. A *blockmodel* is a network which is a structural abstraction of a data network. Sets of nodes in the data network which have similar patterns of ties with all other such sets are represented by single model nodes, called *blocs* (Harrison C. White, Scott A. Boorman, and Ronald L. Breiger, "Social Structure from Multiple Networks, I: Blockmodels of Roles and Positions," *American Journal of Sociology* 81 (1976): 730–80). Blockmodeling will be discussed in detail in Chapter 4.

42. These terms and distinctions are drawn from Berkowitz and Heil, op. cit., 1980.

43. Karl Marx, *A Contribution to the Critique of Political Economy* (Moscow: Progress Publishers, n.d.), especially the preface; Karl Marx, *Capital* (Moscow: Progress Publishers, n.d.), pp. 66–72 (on the labor theory of value) and pp. 543–600 (on surplus value and social capital). Richard and Fernande De George, op. cit., 1972, pp. xii-xvi, contains an excellent short discussion of Marx's contributions to structuralism. Henri de St.-Simon, *Social Organization, the Science of Man*, ed. and trans. Felix Markham (New York: Harper & Row, 1964), pp. 78–116. On social contract theory, see Jean-Jacques Rousseau, *The Social Contract and Discourses* (London: Dent, 1973), pp. 165–319. G. D. H. Cole's introduction to the edition is an excellent summary not only of Rousseau's writings, but of social contract theory and its historical context generally. Despite differences, note the similarity between Rousseau and Hobbes on the question of individualism (Thomas Hobbes, *De Cive* or *The Citizen* [New York: Appleton-Century-Crofts, 1949], pp. 21-69).

44. In his "Letters from an Inhabitant of Geneva," St.-Simon distinguishes three classes which constitute different "sections of humanity": "scientists, artists and men of liberal ideas," all other property holders, and "the rest of humanity" (St.-Simon, op. cit., 1964, p. 2). His "Letters. . ." address each of the groups separately, appealing, in each case, to their distinct, economically conditioned interests. Marx and, especially, Engels were particularly taken with these passages because of their implicit recognition of the

underlying *economic* basis for social perceptions and action (Frederich Engels, *Anti-Dühring* [Moscow: Progress Publishers, 1969], p. 307). This notion that "man" or "humanity" contained distinct subgroups which could be distinguished from one another on the basis of their—to choose Marx's phrase—"social relations of production" was one of the great revelations of nineteenth-century thought.

45. Marx's early writings used the term *man* or *men* in referring to social actors. Indeed, even when he spoke of "classes," these were conceived of as "classes of men," i.e., as sets of individuals (Karl Marx, *Economic and Philosophic Manuscripts of 1844* [Moscow: Foreign Languages Pub., 1961]). In his later works—and, particularly, in *Capital*—classes, themselves, become collective social actors. This change in terminology reflected a growing recognition of the importance of levels of structure in political economy.

46. Emile Durkheim, *The Rules of Sociological Method* (London: Free Press, 1964b), pp. 1–13.

47. S. F. Nadel recognized that the problem of levels was intrinsic to the modeling process, i.e., that there was no inherently social level of facts apart from the *degree of abstraction* at which analysts constructed their models (S. F. Nadel, *The Theory of Social Structure* [Glencoe, Ill.: Free Press, 1957], pp. 97–124). Units of analysis, i.e., the primary units used in explaining a phenomenon, are, as a result, crucial: units constructed at inappropriate levels of abstraction yield misleading results. Nadel was almost alone among social scientists in his generation in realizing that the reason one could not simply "build up" more complex social structures from simpler ones was not that there was a fundamentally irreducible set of "social phenomena" which "emerged" synergistically at certain "levels" of a system, but that abstraction was needed in order to make the "qualitative content" of observations commensurable (ibid., pp. 106–8). The more orthodox view is presented by Peter Blau: "Although the division of labor in a community refers ultimately to observable patterns of conduct of individuals, it is an emergent property of communities that has no counterpart in a corresponding property of individuals. Age distribution, similarly, is an attribute that exists only on the group level; individuals have no age distribution, only an age" (Peter M. Blau, *Exchange and Power in Social Life* [New York: Wiley, 1964], p. 3).

48. A "household" is almost always defined by census bureaus as a group of related persons living under one roof. "Household income" or "family income," "household size," "household composition" and other derivative concepts are then constructed from it. Young and Willmott and others, however, have discovered empirically that these types of definitions are less than adequate: "People live together and eat together—they are considered to be in the same household. But what if they spend a good part of the day and eat (or at least drink tea) regularly in someone else's household? The households are then to some extent merged. This is most obvious where two families actually live in the same house" (Michael Young and Peter Willmott, *Family and Kinship in East London* [London: Penguin, 1962], p. 47). These variations in household structure are even more important where there is extreme pressure on the housing stock—and hence a variety of housing arrangements—but ties among families and family-based decision making remains strong (Janet Salaff, *Working Daughters of Hong Kong: Filial Piety or Power in the Family?* [Cambridge and New York: Cambridge University Press, 1981]).

49. Nadel first recognized the value of structural analytic methods in bringing about commensurability in data (Nadel, op. cit., 1957). In Chapter 3, we will examine a concrete application of these techniques to the problem of defining effective decision-making units in a corporate structure.

50. A "social group" is a set of persons sharing a given attribute. Thus, if some set of persons share common origins, we may define them as an ethnic group, e.g., "Italian-Americans," "Armenians." Similarly, if they share common life-situations, we may define them, in the Weberian sense, as a "social class" or "status group." Usually there is the suggestion that "social groups" are more than simple sets—that they are joined to one another through ties. This property can be accommodated within this framework by specifying that a "tie" constitutes a shared attribute. However, this apparently easy procedure may introduce methodological difficulties (S. D. Berkowitz, "Structural and Non-structural Models of Elites," *Canadian Journal of Sociology* 5 (1980): 13–30). At present, at least three kinds of "sharing" are reflected in definitions of social groups: (a) sharing of simple attributes, e.g., income, common origins; (b) sharing of direct relations to one another, e.g., friendship; and (c) sharing common relationships to others, e.g., relationships to subordinates. Both of the latter two senses can be considered exclusively structuralist.

51. See Georg Simmel, "The Metropolis and Mental Life," in Kurt H. Wolff (ed. and trans.), *The Sociology of Georg Simmel* (Glencoe, Ill.: Free Press, 1950), pp. 407–24. Also, Kadushin, op. cit., 1966.

52. In aggregative analysis, "ordering" simply implies a classification or ranking of elements according to some attribute—as, for instance, when we arrange the students in a class into a rank order on the basis of their test scores. They could be "ordered," in a similar fashion, on the basis of any attribute which discriminated among members of the set. Each element, however, is independent of the others: the "score" on a given attribute which is used in ordering the set is not dependent on the score of any other element. Structuralists prefer orderings of elements which reflect *relations* between them. While Marx himself contributed in important ways to the development of structuralism (see notes 1, 43, 44, and 45 of this chapter), most Marxists today simply "borrow" data and implicit conceptualizations from aggregative analysts. Thus, in this sense, their conception of "ordering" tends to be quite conventional.

53. M. Glanzer and R. Glaser, "Techniques for the Study of Group Structure and Behavior: I. Analysis of Structure," *Psychological Bulletin* 56 (1959): 317–32; A. L. Epstein, "The Network and Urban Social Organization," *Rhodes-Livingston Journal* 29 (1961): 29–62. In order to achieve this "internal coherence and relative external isolation" a cluster must, obviously, be quite dense, and weakly connected to others. Thus, it is a form of subgraph whose limiting case is a *clique*, i.e., a perfectly dense, completely isolated subgraph. Mitchell defined a "cluster" as "a set of (elements) whose links to one another are comparatively dense without necessarily constituting a clique in the strict sense" (Mitchell, op. cit., 1969, p. 64). Also see James A. Davis, "Clustering and Balance Theory in Graphs", *Human Relations* 20 (1967): 181–87.

54. Frank Harary, "Graph Theoretic Measures in the Management Sciences," *Management Science* 5 (1959): 387–403. Since we define a *clique* in terms of adjacency, i.e., its members are 1-distant from one another, there is

172 Notes to Pages 15-17

obviously a relationship between clustering (the weak form) and cliquing (the stronger form) and the path-distance between nodes. In general, the further nodes are from one another, the less likely they are to fall into the same cluster according to *any* clustering measure. See Patrick Doreian, "A Note on the Detection of Cliques in Valued Graphs," *Sociometry* 32 (1969): 237-42.

55. Important examples of research which utilize clustering techniques will be described later. Also see R. N. Shepard and Phipps Arabie, "Additive Clustering: Representation of Similarities as Combinations of Discrete Overlapping Properties," *Psychological Review* 86 (1979): 87-123, for a method of clustering which permits controlled overlapping of clusters (sometimes called "clumps"). James S. Coleman and D. MacRae ("Electronic Data Processing of Sociometric Data for Groups up to 1000 in Size," *American Sociological Review* 25 (1960): 722-26) explore the relationship between clusters and structures created by strong ties which fall outside cluster partitions.

56. Lorrain and White, op. cit., 1971, contains the clearest statement of the precise relationships among these concepts.

57. See Helen Jennings, "Structure of Leadership: Development and Sphere of Influence," *Sociometry* 1 (1937): 99-143; M. Gurevitch, "The Social Structure of Acquaintanceship Networks" (Ph.D. dissertation, Massachusetts Institute of Technology, Cambridge, Mass., 1961); M. Gurevitch and A. Weingrod, "Who Knows Whom: Contact Networks in the Israeli National Elite," *Megamot* 22 (1976): 357-78; C. Lundberg, "Patterns of Acquaintanceship in Society and Complex Organization: A Comparative Study of the Small World Problem," *Pacific Sociological Review* 18 (1975): 206-22.

58. Ithiel de Sola Pool and Manfred Kochen, "Contacts and Influence," *Social Networks* 1 (1978/79): 5-51.

59. Stanley Milgram, "The Small World Problem," *Psychology Today* 1 (1967): 61-67; Stanley Milgram, "Interdisciplinary Thinking and the Small World Problem," in M. Sherif and C. Sherif (eds.), *Interdisciplinary Relationships in the Social Sciences* (Chicago: Aldine, 1969), pp. 103-20; C. Korte and S. Milgram, "Acquaintanceship Networks between Racial Groups: Application of the Small World Method," *Journal of Personality and Social Psychology* 15 (1970): 101-8; J. Travers and S. Milgram, "An Experimental Study of the Small World Problem," *Sociometry* 32 (1969): 425-43. The article by de Sola Pool and Kochen, cited in note 58, above, was actually written some 20 years earlier and has been circulating in manuscript form.

60. Harrison C. White, "Search Parameters for the Small World Problem," *Social Forces* 49 (1970a): 259-64; Harrison C. White, "Everyday Life in Stochastic Networks," *Sociological Inquiry* 43 (1973): 43-49.

61. Mark Granovetter, "The Strength of Weak Ties," *American Journal of Sociology* 78 (1973): 1360-80; Mark Granovetter, *Getting a Job* (Cambridge, Mass.: Harvard University Press, 1974); Nancy Howell (Lee), *The Search for an Abortionist* (Chicago: University of Chicago Press, 1969); Barry Wellman, "Urban Connections" (Toronto: Centre for Urban and Community Studies, University of Toronto, 1976); Paul Craven and Barry Wellman, "The Network City," *Sociological Inquiry* 43 (1973): 57-88; Barry Wellman, "The Community Question: The Intimate Networks of East Yorkers," *American*

Journal of Sociology 84 (1979): 1201–31; Scott A. Boorman, "A Combinatorial Optimization Model for Transmission of Job Information through Contact Networks," *Bell Journal of Economics* 6 (1975): 216–49.

62. On the properties of highly sparse networks, see Jeremy F. Boissevain, *Friends of Friends* (Oxford: Basil Blackwell, 1974); Robert L. Walker, "Social and Spatial Constraints in the Development and Functioning of Social Networks: A Case Study of Guildford (Ph.D. dissertation, London School of Economics, 1974).

63. *Attributes* are characteristic properties of elements which can be used to discriminate them from one another (see note 50). In *aggregative analysis*, researchers begin by defining a series of attributes onto a set of such units, i.e., Where A is a given attribute, then A_{r_i} is the value of that attribute associated with element x_i. These values are then "aggregated" together to provide summary statistics on each defined attribute. See Berkowitz, op. cit., 1980, and Chapter 2 for a more complete discussion of the differences between aggregative and structural modeling.

64. For an excellent, concise review of the literature in the area, see Linton C. Freeman, "Centrality in Social Networks: Conceptual Clarification," *Social Networks* 1 (1978/79): 215–39.

65. Linton C. Freeman, op. cit., 1978/79, p. 221.

66. As cited in Freeman, op. cit., 1978/79, Nieminen measures the degree-centrality, C_D, of a given point, P_k, as

$$C_D(P_k) = \sum_{i=1}^{n} a\,(P_i, P_k)$$

where $a\,(P_i, P_k) = 1$ iff P_i and P_k are connected, 0 otherwise.

Therefore, *relative centrality*, i.e., the measure of centrality generated when sheer size of network is disregarded, C'_D, is given by

$$C'_D\,(P_k) = \frac{\sum_{i=1}^{n} a\,(P_i, P_k)}{n-1}$$

Note here that "opportunities for contact between nodes" are assumed to reflect the "maximum" number of possible ties under conditions where no constraint is present $(n-1)$. Implicit in this is some notion of "cost": that the number of ties at point P_k would reach its maximum if no gradient of costs extrinsic to the network existed at that point.

67. Freeman, op. cit., 1978/79, pp. 221–26; Harary et al., op. cit., 1965, pp. 134–41.

68. Freeman, op. cit., 1978/79; Murray Beauchamp, "An Improved Index of Centrality," *Behavioral Science* 10 (1965):161–63.

69. Linton C. Freeman, "A Set of Measures of Centrality Based on Between-ness," *Sociometry* 40 (1977): 35–41; Phillip Bonacich, "Technique for

Analyzing Overlapping Memberships," in Herbert L. Costner (ed.), *Sociological Methodology* (San Francisco: Jossey-Bass, 1972), pp. 176–85.

70. These are discussed in Chapter 3.

71. The notion of "detection" is used here because, in most empirically obtained networks, particular substructures are not obvious due to (a) the large number of ties involved, (b) the relatively arbitrary arrangement of nodes in a given graph, and (c) the possibility that larger structures may be contained within smaller ones. In addition, in some cases it is necessary to suppress transient features of a network ("detail") in order to bring out enduring or persistent structure ("outline"). See Berkowitz and Heil, op. cit., 1980.

72. Wolff has rendered Simmel's term, "Verbindung zu dreien"—literally, "association of three"—as *triad*. While the terms are not literally the same, Simmel intended substantially the meaning that we attach to them today. Similarly, "Zweierverbindung" has been rendered as *dyad*. Wolff (trans. and ed.), op. cit., 1950, pp. 122–42.

73. Fritz Heider, "Attitudes and Cognitive Organization," *Journal of Psychology* 21 (1946):107–12.

74. Cartwright and Harary, op. cit., 1956.

75. Davis, op. cit., 1967.

76. James A. Davis and Samuel Leinhardt, "The Structure of Positive Interpersonal Relations in Small Groups," in Joseph Berger, Morris Zelditch, Jr., and Bo Anderson (eds.), *Sociological Theories in Progress*, 2 (Boston: Houghton Mifflin, 1971).

77. Paul Holland and Samuel Leinhardt, "A Unified Treatment of Some Structural Models for Sociometric Data," Pittsburgh: Carnegie Mellon University, Technical Report, 1970a.

78. Paul Holland and Samuel Leinhardt, "A Method for Detecting Structure in Sociometric Data," *American Journal of Sociology* 70 (1970b):492–513.

79. S. D. Berkowitz, Peter J. Carrington, Yehuda Kotowitz, and Leonard Waverman, "The Determination of Enterprise Groupings through Combined Ownership and Directorship Ties," *Social Networks* 1 (1978/79):391–413.

80. A Gini index of 0 reflects perfect equality and of 1, perfect inequality.

2: Kin, Friends, and Community

1. This so-called "rule of reasonability" is embodied in the English common law, and, hence, its influence on various aspects of law and jurisprudence is pervasive. Sociologists have primarily been interested in it as an aspect of the interface between customary practice and law (Phillip Selznick, "Sociology and Natural Law," *Natural Law Forum* 6 (1964): 84–108), as it affects crime and definitions of individual responsibility (Jack Gibbs, *Crime, Punishment, and Deterence* [New York: Elsevier, 1975]), and its relationship to definitions of mental illness (Marie Jahoda, *Current Concepts of Positive Mental Health* [New York: Basic, 1958]). The trend among sociologists is to reject the assumption that there is an implicit relationship between

individual behavior and *either* "reason" or "unreason" but, instead, to look at how institutional processes shape definitions of conduct (Michel Foucault, *Madness and Civilization: A History of Insanity in the Age of Reason* [New York: Pantheon, 1965]), the functions of law and legal institutions (Talcott Parsons, "The Law and Social Control," in William E. Evan (ed.), *Law and Society* [New York: Free Press, 1962]), or law as a form of social behavior (Donald Black, "The Boundaries of Legal Sociology," *Yale Law Journal* 81 (1972): 1086–100). Traditionalists, however, continue to adhere to the rationalist view of law (H. L. A. Hart, *The Concept of Law* [Oxford: Clarendon Press, 1961]).

2. As a working assumption for methodological or modeling purposes, these presumptions are incorporated into game theory (John Von Neumann and Oskar Morgenstern, *Theory of Games and Economic Behavior* [New York: Wiley, 1944], pp. 8–15), into economics proper, and into a range of models of like situations (Anatol Rapoport, *Fights, Games, and Debates* [Ann Arbor: University of Michigan Press, 1974]). This is quite different, however, from models of social *reality* which assume that individual behavior is actually shaped by rational calculation or action—as in much of social psychology (Tamatsu Shibutani, *Society and Personality: An Interaction Approach to Social Psychology* [Englewood Cliffs, N.J.: Prentice-Hall, 1961])—or the complementary notion that social *dysfunction* results from the absence of rationality (Kimball Young, *Personality and Problems of Adjustment* [New York: Appleton-Century-Crofts, 1947]). I argue that even some sociologists who see individuals as largely a product of their social environment have adopted a fundamentally individualist perspective, e.g., George H. Mead (*Mind, Self, and Society* [Chicago: University of Chicago Press, 1934]).

3. If this were not so, it would be impossible for psychoanalysts to *interpret* the unconscious' outcroppings, as in Freud's treatment of the relationship between ego and id (Sigmund Freud, *The Ego and the Id*, trans. Joan Riviere [London: Hogarth Press, 1950]) and in his analysis of anxiety (Sigmund Freud, *The Problem of Anxiety* [New York: W. W. Norton, 1936]).

4. Von Neumann and Morgenstern, op. cit., 1944, is probably the clearest statement of the postulate of the "rational optimizer." Economists, per se, usually simply assume this construct.

5. See Ronald Rohner, *The Ethnography of Franz Boas* (Chicago: University of Chicago Press, 1969). Rohner makes the point that the potlatch had already been considerably altered by the introduction of outside sources of wealth by the time Boas studied it in the 1880s.

6. See Anthony Leeds, "The Culture of Poverty: Conceptual, Logical, and Empirical Problems with Perspectives from Brazil and Peru," in E. Leacock (ed.), *The Culture of Poverty: A Critique* (New York: Simon and Schuster, 1970).

7. See Harold Christensen (ed.), *Handbook of Marriage and the Family* (Chicago: Rand McNally, 1964). In addition to traditional perspectives, the *Handbook* also incorporates some excellent articles which challenge the established wisdom. In particular, see Panos D. Bardos, "Family Forms and Variations Historically Considered." Also see, Constantina Safilios-Rothschild, "Towards a Cross-cultural Conception of Family Modernity," *Journal of Comparative Family Studies* 1 (1970): 17–25.

8. See Shibutani, op. cit., 1961.

9. Elizabeth Bott, *Family and Social Network: Roles, Norms, and External Relationships in Ordinary Urban Families* (London: Tavistock, 1957).

10. Ibid., pp. 17–30.

11. Ibid., pp. 35–39.

12. See J. Clyde Mitchell (ed.), *Social Networks in Urban Situations: Analyses of Personal Relationships in Central African Towns* (Manchester: Manchester University Press, 1969).

13. The term *cliquishness* is used here in preference to Bott's *connectedness*, which has a different mathematical interpretation. See Bott, op. cit., 1957, p. 61ff. By "close-knit" and loose-knit" Bott really means the propensity to fall into cliques or clusters; hence my term.

14. Bott, op. cit., 1957, p. 60.

15. Ibid., pp. 61–96.

16. Michael Young and Peter Willmott, *Family and Kinship in East London* (London: Penguin, 1962).

17. Ibid., pp. 31–43.

18. Ibid.

19. Ibid., p. 77.

20. Ibid., p. 131.

21. Ibid., pp. 134–136.

22. Michael Young and Peter Willmott, *The Symmetrical Family* (New York: Pantheon, 1973).

23. See Leslie Howard, "Industrialization and Community in Chotangpur" (Ph.D. dissertation, Harvard University, Cambridge, Mass., 1974). The tightest and most theoretically sophisticated formulation of chain effects is presented in Harrison C. White, *Chains of Opportunity: System Models of Mobility in Organizations* (Cambridge, Mass.: Harvard University Press, 1970b).

24. John A. Barnes, "Class and Committees in a Norwegian Island Parish," *Human Relations* 7 (1954): 39–58.

25. Adrian Mayer, "The Significance of Quasi-Groups in the Study of Complex Societies," in M. Banton (ed.), *The Social Anthropology of Complex Societies* (London: Tavistock, 1966).

26. Bruce Kapferer, "Norms and the Manipulation of Relationships in a Work Context," in Mitchell, op. cit., 1969, pp. 181–244.

27. Prudence Wheeldon, "The Operation of Voluntary Associations and Personal Networks in the Political Processes of an Inter-ethnic Community," in Mitchell, op. cit., 1969, pp. 128–80.

28. Peter Harries-Jones, " 'Home-boy' Ties and Political Organization in a Copperbelt Township," in Mitchell, op. cit., 1969, pp. 297–347.

29. David Boswell, "Personal Crises and the Mobilization of the Social Network," in Mitchell, op. cit., 1969, pp. 245–96.

30. In particular to the work of Frank Harary and his collaborators, which he heavily cites. Mitchell, op. cit., 1969, pp. 1–50.

31. Frank Harary, Robert Z. Norman, and Dorwin Cartwright, *Structural Models: An Introduction to the Theory of Directed Graphs* (New York: John Wiley & Sons, 1965), pp. 32–40.

32. See S. D. Berkowitz, "The Dynamics of Elite Structure: A Critique of C. Wright Mills' 'Power Elite' Model" (Ph.D. dissertation, Brandeis University, Waltham, Mass., 1975).

33. See Mitchell, op. cit., 1969, pp. 1–50, for practical examples of these calculations.

34. Ronald L. Breiger, "The Duality of Persons and Groups," *Social Forces* 53 (1974): 181–90.

35. See note 13.

36. See Emile Durkheim, *The Division of Labor in Society* (New York: Free Press, 1964a); Emile Durkheim, *Moral Education* (New York: Free Press, 1961).

37. Thomas Kuhn, *The Structure of Scientific Revolutions* (Chicago: University of Chicago Press, 1970). We will examine this question in detail in Chapter 5.

38. S. D. Berkowitz, "Structural and Non-structural Models of Elites," *Canadian Journal of Sociology* 5 (1980): 13–30. Much of the argument here is condensed from this source.

39. Berkowitz, op. cit., 1980. The technical means of doing this, in an explicit way, will become clear in Chapters 3 and 4 when we deal with algebraic modeling.

40. Berkowitz, op. cit., 1980.

41. Harrison C. White, *An Anatomy of Kinship* (Englewood Cliffs, N.J.: Prentice-Hall, 1963).

42. Ibid., pp. 6–7.

43. Ibid., p. 8.

44. Ibid., pp. 9–10.

45. Ibid., p. 10.

46. Ibid.

47. Ibid., pp. 13–22.

48. Ibid., p. 27.

49. A "permutation matrix" is a square matrix in which exactly one "1" appears in each row and column and all other cells are zero-filled.

50. H. C. White, op. cit., 1963, p. 35.

51. Ibid.

52. Ibid., pp. 36–38.

53. See John Paul Boyd, "The Algebra of Group Kinship," *Journal of Mathematical Psychology* 6 (1969): 139–67.

54. Algebraic techniques are particularly appropriate for doing this because they force researchers to deal with (a) whole-part relationships among elements and (b) analytically distinct *dimensions* of social structure.

55. H. C. White, op. cit., 1963, pp. 94–149.

56. Western kinship systems cannot be easily dealt with in White's terms because (a) they are not divided into mutually exclusive and exhaustive "sections," as the societies White studied are, and (b) they do not exhibit clear prescriptive marriage rules. In principle, however, despite the fact that they are far more complex than aboriginal groups, Western societies do exhibit *relative* segregation in social contacts and marriage choice and, therefore, could be amenable to modified, and less structured, analyses of the type White performed.

57. H. C. White, op. cit., 1963, pp. 5–6.

58. Probably the prevailing tendency among social scientists at the time would have been to view structural analysis as a collection of very difficult formal tools aimed at coping with precisely the same kinds of issues that motivate "standard" social scientists. This perception would have been due, in part, to the fact that structuralists published their findings in journals which were to a large extent empirically oriented. Thus, often only their substantive findings "stuck" in readers' minds. Structured opportunities for publication of purely theoretical work were, until the late 1960s, few and far between.

59. A "solidary" community is one based on strong personal and direct ties.

60. See Karl Marx, *A Contribution to the Critique of Political Economy* (Moscow: Progress Publishers, n.d.); Karl Marx, *Selected Works* (New York: International Publishers, n.d.), pp. 204–28, 356–75; Hans Gerth and C. Wright Mills, *From Max Weber: Essays in Sociology* (New York: Oxford, 1946), pp. 196–244; Durkheim, op. cit., 1964a, pp. 111–32; Georg Simmel, "The Metropolis and Mental Life," in Kurt H. Wolff (ed. and trans.), *The Sociology of Georg Simmel* (Glencoe, Ill.: Free Press, 1950), pp. 407–24.

61. See note 60.

62. Barry Wellman, "The Community Question: The Intimate Networks of East Yorkers," *American Journal of Sociology* 84 (1979): 1201–31.

63. Maurice Stein, *The Eclipse of Community* (Princeton, N.J.: Princeton University Press, 1960).

64. C. Wright Mills, *White Collar: The American Middle Classes* (New York: Oxford, 1951); William Kornhauser, *The Politics of Mass Society* (Glencoe, Ill.: Free Press, 1959).

65. Wellman, op. cit., 1979.

66. Herbert Gans, *The Urban Villagers: Group and Class in the Life of Italian Americans* (New York: Free Press, 1962).

67. Wellman, op. cit., 1979.

68. Ibid.

69. Edward O. Laumann, *Bonds of Pluralism: The Form and Substance of Urban Social Networks* (New York: Wiley, 1973), pp. ix–xi.

70. Ibid, pp. 17–21.

71. Claude S. Fischer, Robert Max Jackson, C. Ann Stueve, Kathleen Gerson, and Lynne McCallister Jones, with Mark Baldassare, *Networks and Places: Social Relations in the Urban Setting* (New York: Free Press, 1977).

72. Ibid., p. 11.

73. Ibid., pp. 8, 11.

74. Ibid.

75. Ibid., p. 6.

76. Ibid., p. 6.

77. See Rapoport, op. cit., 1974.

78. In order to be interpretable in a game theoretic sense, a "choice" must be made between options whose *potential* consequences are known. As Fischer et al. present their model, this is not the case (see Rapoport, op. cit., 1974, pp. 14–50).

79. See Morris Hamburg, *Statistical Analysis for Decision Making*, 2nd ed. (New York: Harcourt, Brace, Jovanovich, 1977), for a systematic review of the methods involved.

80. "Forecasting" is the use of formal models to establish the likelihood of future outcomes and the costs and benefits of alternative courses of action associated with them. See Hamburg, op. cit., 1977.

81. In this procedure, all two-way combinations of nodes are considered in isolation from all others, i.e., they are treated as though no combinations ≥ 3 exist.

82. Fischer et al., op. cit., 1977, pp. 39–98, contains a summary of the methods used in this study.

83. Laumann, op. cit., 1973, pp. 2–8.

84. Ibid., pp. 10–11, 73–74.

85. Ibid., p. 75.

86. Ibid., pp. 83–110.

87. The term *mini-network*, as used here, refers to a network anchored on ego (sometimes called a reticulum) which has then been "closed" by asking alters about their relationships with one another. This is fundamentally the same sort of graph generated in Bott's study. It is intended as some sort of *index* of the structural properties of the hypothetical network from which it has been drawn—hence the term. However, it is clearly *not* (a) a simple reticulum, (b) a random sample taken from all possible graphs that *could* be

drawn on the included points, or (c) a subgraph that is morphologically representative of the whole.

88. Laumann, op. cit., 1973, pp. 67, 81–8.

89. Ibid., pp. 98–105.

90. See Wellman, op. cit., 1979, pp. 1204–5.

91. Laumann, op. cit., 1973, 27–132, passim.

92. This observation—and a number of other important points about the historic use of spatial models in sociology—is made in David D. McFarland and Daniel J. Brown, "Social Distance as a Metric: A Systematic Introduction to Smallest Space Analysis," which is included as a chapter in Laumann, op. cit., 1973, pp. 213–53.

93. Georg Simmel, "The Stranger," in Wolff, op. cit., 1950, pp. 402–8.

94. See Robert E. L. Faris, *Chicago Sociology: 1920–1932* (San Francisco: Chandler, 1967).

95. Pitrim A. Sorokin, as cited in McFarland and Brown, op. cit., 1973, pp. 214–17.

96. McFarland and Brown, op. cit., 1973, pp. 219–22.

97. When two metrics with very different properties are combined without recognizing these differences, the combined metric often has unusual characteristics. See H. L. Royden, *Real Analysis* (New York: Macmillan, 1963).

98. Ithiel de Sola Pool and Manfred Kochen, "Contacts and Influence," *Social Networks* 1 (1978/79): 5–51.

99. M. Gurevitch, "The Social Structure of Acquaintanceship Networks" (Ph. D. dissertation, Massachusetts Institute of Technology, Cambridge, Mass., 1961).

100. See Nan Lin and Paul W. Dayton, "The Urban Communication Network and Social Stratification: A Small World Experiment" (paper presented at the annual meetings of the International Communication Association, Portland, Oregon, 1976).

101. J. Travers and S. Milgram, "An Experimental Study of the Small World Problem," *Sociometry* 32 (1969): 425–43.

102. This observation is the basis of the "liberated" community postulated in Wellman, op. cit., 1979.

103. C. Korte and S. Milgram, "Acquaintanceship Networks between Racial Groups: Application of the Small World Method," *Journal of Personality and Social Psychology* 15 (1970): 101–8 at 102.

104. Ibid.

105. Ibid., p. 103.

106. Ibid., pp. 103–6.

107. If one accepts the definition of *power* as the ability to mobilize a system of relations toward given ends, then structural discontinuities represent limits on the power of a group which are intrinsic to its pattern of communication. In terms of the classic "small world" approach, if members of a group cannot easily "contact" decision makers or sources of information, that group as a whole is handicapped in making use of these resources. In addition, since the time when the main body of small world studies was conducted, we have learned a great deal about the importance of relatively short communications paths in the formation of group consensus, e.g., in studies of density of ties. Gross discontinuities in a communications pattern, therefore, obviously inhibit the formulation of plans of action which will enjoy widespread support from a group. The strong suggestion is made, inter alia, in Korte and Milgram's study that these factors were at work among blacks—relative to whites—at the time their study was done. Since a suggestion is not the same thing as a conclusion, this is clearly a topic that should be addressed more directly and rigorously. Moreover, given the changes that have taken place in the intervening years—in particular, the higher proportion of blacks in the professions—it would be quite possible and interesting to examine whether or not the parameters observed in Korte and Milgram's work have altered significantly. Finally, the theoretical and methodological questions raised in Korte and Milgram should be tackled directly: What *precisely* is the relationship between path-distances and the mobilization of resources in *very* large groups? Can we learn something from this about the practical mechanics of "stratification" in the general case? And, under controlled conditions, how do differences in power become institutionalized in a social structure?

108. Lin and Dayton, op. cit., 1976.

109. René Thom, *Structural Stability and Morphogenesis,* trans. D. H. Fowler (Reading, Mass.: W. A. Benjamin, 1975).

110. The most immediate applications of catastrophe theory in social sciences appear to be to studies of human and animal behavior. However, in principle, it ought to be of interest wherever discontinuous or divergent phenomena have been observed. Thom identifies seven basic forms of catastrophes—each corresponding to a different topological representation: "fold," "cusp," "swallowtail," "butterfly," "hyperbolic," "elliptic," and "parabolic" (see Thom, op. cit., 1975).

111. Wellman, op. cit., 1979, pp. 1209–10.

112. Ibid., pp. 1211–14.

3: Corporations and Privilege

1. S. D. Berkowitz, "The Dynamics of Elite Structure: A Critique of C. Wright Mills' 'Power Elite' Model" (Ph.D. dissertation, Brandeis University, Waltham, Mass., 1975), pp. 212–360; John P. Davis, *Corporations* (New York: Capricorn, 1961); Samuel Richardson Reid, *The New Industrial Order: Concentration, Regulation, and Public Policy* (New York: McGraw-Hill, 1976).

2. Reid, op. cit., 1976, pp. 30–39; Joe S. Bain, *Industrial Organization* (New York: John Wiley & Sons, 1968), pp. 112–55; Charles Levinson, *Capital, Inflation, and the Multinationals* (Winchester, Mass.: Allen & Unwin, 1971), pp. 166–84; M. A. Utton, *Industrial Concentration* (Middlesex: Penguin, 1970).

3. A *multinational corporation* is a firm which is chartered in and operates subsidiaries in more than one national jurisdiction. The head offices of most multinational corporations are in Europe or the United States—although Japanese multinationals are becoming increasingly important (see Levinson, op. cit., 1971). The largest multinationals account for sales in excess of $30 billion. In 1974, the 500 largest industrials averaged $1.1 billion in assets. The 50 largest banks averaged ten times as much (see Reid, op. cit., 1976, p. 27).

4. See Erwin O. Smigel, *The Wall Street Lawyer* (Bloomington: Indiana University Press, 1969), for a study of the largest noncorporate law offices. Also, Roger B. Siddall, *A Survey of Large Law Firms in the United States* (New York: Vantage, 1956).

5. This observation has led to the conclusion, on the part of some analysts, that modern, large-scale corporations function like quasi governments. See Adolph A. Berle and Gardiner C. Means, *The Modern Corporation and Private Property* (New York: Macmillan, 1932).

6. Classic instances of this are provided by the large Japanese multinational corporations called *zaibatsu*.

7. Berkowitz, op. cit., 1975, contains a bibliography which includes most of the significant studies published before it was defended.

8. These are too extensive to be cited here. However, the "case study" method has been a staple of most graduate schools of business administration for some time, and monographs and texts summarizing them are commonplace. See C. A. Heiss, *Accounting in the Administration of Large Business Enterprises* (Cambridge, Mass.: Harvard University Press, 1943); B. B. Howard and M. Upton, *Introduction to Business Finance* (New York: McGraw-Hill, 1953); R. A. Gordon, *Business Leadership in the Large Corporation* (Washington: Brookings Institution, 1943).

9. See *Standard & Poor's Register of Corporations, Directors, and Executives* (New York: Standard & Poor's Corporation [annual]); *Moody's Banking and Finance Manual* (New York: Moody's Investors' Service [annual]); *Moody's Industrial Manual* (New York: Moody's Investors' Service [annual]); *Financial Post, Survey of Industrials* (Toronto: Maclean Hunter [annual]); etc. There are literally hundreds of similar sources.

10. During the course of the study reported later in this chapter, the Toronto group found that between four and seven percent of entries in *different* private sources did not coincide. The error rates were substantially higher in one government source. See S. D. Berkowitz, Yehuda Kotowitz, and Leonard Waverman, with Bruce Becker, Randy Bradford, Peter Carrington, June Corman, and Gregory Heil, *Enterprise Structure and Corporate Concentration* (Ottawa: Royal Commission on Corporate Concentration, 1976 [issued 1978]); S. D. Berkowitz, P. J. Carrington, J. S. Corman, and L. Waverman, "Flexible Design for a Large-scale Corporate Data Base," *Social Networks* 2 (1979): 75–83.

11. I use the term *unstructured* here because conventional economics assumes that the only effects one participant has on another are indirect and mediated through their common participation in that market. This may provide a common structure only if there is an overarching set of constraints in terms of which bargaining takes place (see S. D. Berkowitz, "Markets and Market-Areas," in S. D. Berkowitz and Barry Wellman (eds.), *Structural Sociology* (Cambridge and New York: Cambridge University Press, forthcoming).

12. This argument—and hence the definition of a firm—is confined to private business concerns. However, *state*-operated business concerns can, under certain circumstances, be treated as firms as well.

13. Defined functionally, firms are units of optimization in decision making. Legally defined, they consist of a group of assets held in a common name. See the discussion of the "Toronto group" later in this chapter for a more complete elaboration of the difference between the two usages.

14. For the most complete and straightforward statement of the paradigm, see Richard Caves, *American Industry: Structure, Conduct, Performance*, 4th ed. (Englewood Cliffs, N.J.: Prentice-Hall, 1977).

15. See Chapter 2 for a discussion of asymmetry as a property of graphs. Since graphs are used to model the relationship among elements of systems, the degree of asymmetry in a graphic representation of that system may be used as an index of the degree to which nodes are dominated or controlled by others. See S. D. Berkowitz, P. J. Carrington, Yehuda Kotowitz, and Leonard Waverman, "The Determination of Enterprise Groupings through Combined Ownership and Directorship Ties," *Social Networks* 1 (1978/79): 391–413, and later discussions in this chapter.

16. These factors are referred to as "nonmarket" because they fall outside the conventional definitions of market activity. As we shall see later on, structural models can be constructed where they bear on that activity in important ways.

17. Rudolph Hilferding, *Das Finanzkapital* (1910; reprint ed., Frankfurt am Main: Europäische Verlaganstalt, 1968).

18. In addition to the work of the Stony Brook group, which is described in detail here, see Jean-Marie Chevalier, *La Structure Financiere de l'Industrie Americaine* (Paris: Cujas, 1970); Robert Fitch and Mary Oppenheimer, "Who Rules the Corporations?" *Socialist Revolution* (1970): 73–107 (July/Aug.); 61–114 (Sept./Oct.); 33–94 (Nov./Dec.); James Knowles, "The Rockefeller Financial Group" (Warner Modular Publications, No. 343, 1973); John Lintner, "The Financing of Corporations," in Edward Mason (ed.), *The Corporation in Modern Society* (New York: Atheneum, 1967), pp. 111–201.

19. See Berkowitz et al., op. cit., 1978/79.

20. For an excellent transnational summary of the literature in this area, see John Scott, *Corporations, Classes and Capitalism* (London: Hutchinson, 1979). Michael Useem's "Corporations and the Corporate Elite" (*Annual Review of Sociology* 6 (1980): 41–77) provides a focused and theoretically acute discussion of the role and organization of directors.

21. L. Waverman and R. Baldwin, "Determinants of Interlocking Directorates" (Toronto: Institute for Policy Analysis, 1975), is an especially craftsmanlike piece of work in this genre.

22. A detailed summary of these appears in Meindart Fennema and Huibert Schijf, "Analyzing Interlocking Directorates: Theory and Methods," *Social Networks* 1 (1978/79): 297–332.

23. See U.S. National Resources Committee, *The Structure of the American Economy, Part I: Basic Characteristics* (Washington: U.S. Govt., 1939); U.S. Federal Trade Commission, *Report of the Federal Trade Commission on Interlocking Directorates* (Washington: U.S. Govt. Printing Office, 1951); and Peter C. Dooley, "The Interlocking Directorate," *American Economic Review* 59 (1969): 314–23.

24. C. Wright Mills, *The Power Elite* (New York: Oxford, 1956), p. 123.

25. Ibid.

26. U.S. Federal Trade Commission, op. cit., 1951.

27. In part because it focuses on "directors" rather than clusters of corporations. In this sense, it is a perfect illustration of the duality problem in corporate networks, which we will discuss later in this chapter. Also see S. D. Berkowitz, "Structural and Non-structural Models of Elites," *Canadian Journal of Sociology* 5 (1980): 13–30.

28. Sam Aaronovitch, *The Ruling Class: A Study of British Finance Capital* (London: Lawrence & Wishart, 1961). Also see Sam Aaronovitch, *Monopoly: A Study of British Monopoly Capitalism* (London: Lawrence & Wishart, 1955).

29. Michael Barratt Brown, "The Controllers of British Industry," in K. Coates (ed.), *Can the Workers Run Industry?* (London: Sphere, 1968).

30. For exceptions, see Berkowitz, op. cit., 1980.

31. W. L. Warner and D. B. Unwalla, "The System of Interlocking Directorates," in W. L. Warner, D. B. Unwalla, and J. H. Trimm (eds.), *The Emergent American Society*, vol. 1 (New Haven: Yale University Press, 1967).

32. Dooley, op. cit., 1969.

33. Ibid., pp. 319–20.

34. See Berkowitz, op. cit., 1980.

35. Joel H. Levine, "The Sphere of Influence," *American Sociological Review* 37 (1972): 14–27.

36. Ibid., p. 14.

37. *Nomothetic* means law or lawlike establishing; hence, a model which seeks to uncover lawlike relationships in the "real world." *Idiographic* means image-drawing; hence, a model which seeks to depict something in richness of detail.

38. The term *mapping* here is used in the specific and technical sense we discussed earlier.

39. That is, as a point.

40. See A. P. M. Coxon and Charles Jones, "Multidimensional scaling: Exploration to Confirmation," *Quality and Quantity* 14 (1980): 31–73. Also see, David D. McFarland and Daniel J. Brown, "Social Distance as a Metric: A Systematic Introduction to Smallest Space Analysis," in Edward O. Laumann, *Bonds of Pluralism: The Form and Substance of Urban Social Networks* (New York: Wiley, 1973), pp. 213–53.

41. This means that every node in the subgraph is connected to every other one at some remove.

42. See Coxon and Jones, op. cit., 1980. They argue that MDS-produced scales are "quasi metrics," i.e., that they obey most of the important properties of metrics.

43. The "third dimension" is implied in the separation of the set into banks and industrials.

44. i.e., that of the Stony Brook group, which we will discuss later in this chapter.

45. Levine, op. cit., 1972, p. 20.

46. Ibid., pp. 21–22.

47. Ibid., p. 22.

48. The objects in the map in Figure 3.9 are shown as if they had been projected onto the surface of that ceiling. Just as in a planetarium, however, it is very difficult to tell whether two objects that appear at the same point are or ought to be the same distance from the viewer, or whether one is really farther away. In other words, one such object may be focused at a point on the sphere and the other behind it. The spaces formed by the coordinates—which make objects appear "closer" or "farther away"—are intended to counteract this effect. They cannot, however, entirely eliminate it.

49. These sets are "associated" in the sense that they appear in close proximity to one another within the mapping. In general, firms that share a large number of ties will tend to be closer together and those that share few or none will be further apart. Precisely how close to one another two nodes have to be before this fact is significant, however, still remains to be established.

50. See Harrison C. White, *An Anatomy of Kinship* (Englewood Cliffs, N.J.: Prentice-Hall, 1963).

51. See John Paul Boyd, "The Algebra of Group Kinship," *Journal of Mathematical Psychology* 6 (1969): 139–67.

52. See François Lorrain and Harrison C. White, "Structural Equivalence of Individuals in Social Networks," *Journal of Mathematical Sociology* 1 (1971): 49–80.

53. Here the term *orders of interconnection* refers to products of generators. See Chapter 2.

54. In Chapter 2 we discussed the conditions necessary for *transitivity* to be present. *Intransitivity* is simply the condition where they are not. A graph is

said to be "commutative" if, given G_{sa} and G_{sb} as subgraphs of that graph, it obeys the rules that $G_{sa} \cap G_{sl} = G_{sb} \cap G_{sa}$ and $G_{sa} \cup G_{sb} = G_{sb} \cup G_{sa}$.

55. If there had been anomalous ties, then it would have indicated that our hypothesis, specified in the reduction equation, was in error or, at least, did not perfectly fit the data. This feature of algebraic reductions enables analysts to *falsify* their models.

56. By implication, the distinction between bonded-ties and emanations suggests that we cannot impute dominance to one or another nodes joined by them, i.e, that *all* are mutually bound up within the same set of constraints. Graphic asymmetry or dominance can be built in by compounding bonded ties with other ties.

57. See Lorrain and White, op. cit., 1971.

58. In this sense, Hilferding-like models are ones which attempt to examine the overall architecture of corporate networks as opposed to their specific point-to-point organization. Multiplexity is the property of having more than one tie between the same nodes, i.e., what graph theorists call "parallel arcs."

59. See S. D. Berkowitz and L. Felt, *Phase I of a Structural Analysis of the Canadian Financial System*, 2 vols., Report (Toronto: Institute for Policy Analysis, 1975).

60. See James Bearden, William Atwood, Peter Freitag, Carol Hendricks, Beth Mintz, and Michael Schwartz, "The Nature and Extent of Bank Centrality in Corporate Networks" (paper read at the annual meetings of the American Sociological Association, 1975).

61. Berkowitz, Carrington, Corman, and Waverman, op. cit., 1979.

62. These were not necessary because the Stony Brook group was initially interested in the *architecture* of the pattern of relations between and among corporations and banks, not the impact of this architecture on the behavior of elements.

63. In the case of the corporate network described by Berkowitz, Carrington, Kotowitz, and Waverman (op. cit., 1978/79), 4,101 of 5,306 nodes fell within one connected component, according to one criterion of connectivity. The task, then, was to discriminate among connections and break this cluster down into more meaningful subclusters.

64. Most studies of corporate networks consider only the very largest firms within an economy. I argue that this is like examining the tips of icebergs and drawing conclusions from this about the glacial structure of the entire icepack. In the case of the research reported in Berkowitz, Carrington, Kotowitz, and Waverman (op. cit., 1978/79), we found a substantial number of cases where "peaks" in the structure were unconnected, but other firms farther down the chain of control had weak ties to one another. In fact, much of the research involved distinguishing trivial or accidental interconnections between enterprises from those which reflected some form of control.

65. Berkowitz, Carrington, Kotowitz, and Waverman, op. cit., 1978/79, p. 397.

66. See Anatol Rapoport and William J. Horvath, "A Study of a Large Sociogram," *Behavioral Science* 6 (1961): 279–91.

67. Bain, op. cit., 1968, pp. 4–5: "By an enterprise we refer to a privately owned business firm (owned either by one individual or jointly by several or many . . .), which engages in productive activity of any sort with the opportunity of making a profit. Enterprises defined thus include private firms engaged in manufacturing, wholesale and retail trade, supplying gas and electricity, construction, banking and so forth; they do not include governmentally owned and operated productive operations, nonprofit organizations performing charitable services, and the like."

68. Establishments should not be confused with subsidiaries, which are separately incorporated firms. An *establishment* is the actual physical plant through which goods are produced or services delivered, e.g., "the Mammoth Electric plant in Xerces, Maine."

69. Bain establishes two criteria by which to judge whether or not some set of facilities or assets falls within his definition of an enterprise: (a) legal control of those facilities or assets, by (b) a single management group. Although Bain recognizes that there are a variety of ways in which this "legal control" can be accomplished—and thus includes the term *ultimate* as part of his definition—this simply exacerbates the problem of assignment of particular establishments to enterprises given that (a) neither he, nor any other conventional industrial organization economist, specifies a general definition of what these legal mechanisms are which is sufficiently broad to accommodate a variety of forms of "control," and (b) management groups typically overlap with one another, as we have seen. The root problem here is the failure to recognize a distinction between practical control and legal control, i.e., a distinction, in our terms, between enterprises and firms (see Bain, op. cit., 1968, p. 5).

70. Statistics Canada, for instance, defines enterprises as "all companies owned more than 50% directly or indirectly where 'owned' refers to voting stock" (Statistics Canada, *Industrial Organization and Concentration in the Manufacturing, Mining and Logging Industries* (Ottawa: Information Canada, 1973).

71. Berkowitz, Carrington, Kotowitz, and Waverman, op. cit., 1978/79, pp. 399–400. Frank Harary, Robert Z. Norman, and Dorwin Cartwright, *Structural Models: An Introduction to the Theory of Directed Graphs* (New York: John Wiley & Sons, 1965), pp. 117–23, describes the properties of directed graphs (digraphs) which do not permit cycles (acyclic).

72. The symbol # refers to Boolean arithmetic.

73. Berkowitz, Carrington, Kotowitz, and Waverman, op. cit., 1978/79, p. 399.

74. Effectively, this procedure embodies the idea that a given firm either controls another or it does not, i.e., that, in the general case—and we will deal with the exceptions, joint ventures—there is no such thing as partial control. Thus, assignment of firms to enterprises is done in stages, and control of one firm by another implies *total* control over its activities. Given three firms, A, B, and C, if A controls B, by whatever margin, its control in C is vested in A *as if* A owned all of B. This procedure, of course, follows from the general definition of transitivity provided in Chapter 2.

75. The symbol ## indicates *stepwise transitive closure*.

76. Berkowitz et al., op. cit., 1976 (1978), and Berkowitz, Carrington, Kotowitz, and Waverman, op. cit., 1978/79.

77. The underlying reasoning here is that these arrays ought to reflect robust properties of the corporate structure that is being modeled. Thus, if random perturbations will seriously disturb a set of arrays, it means that minor errors could have produced particular assignments to these enterprises in the first place. By implication, the test being proposed here is a measure of error-proneness.

78. That is to say (a) when we use strength of interconnection as a means of constructing links in binary matrices (e.g., requiring that two ties be made before a "connection" is recognized), assignments do not yield robust arrays, and (b) the resulting patterns are more or less random.

79. Berkowitz, Carrington, Kotowitz, and Waverman, op. cit., 1978/79, p. 409.

80. This is due to two factors: (a) majority control is extremely common in the Canadian case, and (b) many or most minority-control ties are vertical, i.e., between firms in industries that form part of the same production chain, rather than horizontal, i.e., between firms within the same industry. See Berkowitz, op. cit., forthcoming.

81. Beth Mintz, Peter Freitag, Carol Hendricks, and Michael Schwartz, "Problems of Proof in Elite Research," *Social Problems* 23 (1976): 315–24. A *critical experiment* is one which unambiguously distinguishes between theoretically distinct outcomes.

82. Mintz, Freitag, Hendricks, and Schwartz, op. cit., 1976, p. 322.

83. Bearden et al., op. cit., 1975, and Beth Mintz and Michael Schwartz, "The Role of Financial Institutions in Interlock Networks" (paper read at the New Directions in Structural Analysis conference, Toronto, 1978; *American Sociological Review* [forthcoming]).

84. See Levine, op. cit., 1972, for a similar conclusion about the U.S. corporate structure.

85. Implicit here is a definition of power as the ability to mobilize resources toward given ends. This is similar to that proposed in Berkowitz, op. cit., 1975. The Stony Brook group, however, has not formalized its definition in this way.

86. See M. E. Shaw, "Group Structure and the Behavior of Individuals in Small Groups," *Journal of Psychology* 38 (1954): 139–49. D. L. Rogers, "Sociometric Analysis of Interorganizational Relations: Application of Theory and Measurement," *Rural Sociology* 39 (1974): 487–503, summarizes and incorporates much of this early work.

87. Phillip Bonacich, "Technique for Analyzing Overlapping Memberships," in Herbert L. Costner (ed.), *Sociological Methodology, 1972* (San Francisco: Jossey-Bass, 1972), pp. 176–85.

88. Ibid., pp. 178–79.

89. Ibid., pp. 181–84. The technique used in deriving this index simultaneously for all points is similar to and derived from factor analysis.

90. Bearden et al., op. cit., 1975, describes these experiments.

91. The point at issue here is whether or not the square root standardization procedure adequately reflects the degree of diminution in effect of adding additional members to the receiving board. If there were, for instance, a limit to the number of board members who effectively interact on a board, we might want to incorporate this limit into an intensity measure. Similarly, if the impact of adding slots to the receiving board was more gradual than the square root function would suggest, this ought to be reflected as well. Ultimately, of course, it would be desirable for this measure to be associated with some nonarbitrary and empirically observable property of the pattern of corporate connection taken as a whole.

92. See Berkowitz, op. cit., 1980.

93. Mintz and Schwartz, op. cit., 1978 (forthcoming), p. 7.

94. Ibid., p. 12.

95. Ibid., pp. 12–13.

96. Ibid., pp. 12–14.

97. Ibid., p. 14.

98. Ibid., pp. 14–15.

99. Ronald S. Burt, "A Structural Theory of Interlocking Corporate Directorates," *Social Networks* 1 (1978/79): 415–35. Also see a discussion of this point in Peter J. Carrington, "Horizontal Cooptation through Corporate Interlocks" (Ph.D. dissertation, University of Toronto, 1981).

100. Hilferding, op. cit., 1968. Since the number and types of financial intermediaries have increased radically since Hilferding's time, it seems plausible that it is now necessary to enlarge this hypothesis to incorporate divergences among these. Some work has been done which, presumably, could lead to a further specification and test of more sophisticated theories derived from Hilferding's earlier work. See Ronald S. Burt, "Disaggregating the Effect on Profits in Manufacturing Industries of Having Imperfectly Competitive Consumers and Suppliers," *Social Science Research* 8 (1979): 120–43.

101. John A. Sonquist and Thomas Koenig, "Interlocking Directorates in the Top U.S. Corporations: A Graph Theory Approach," *Insurgent Sociologist* 5 (1976): 196–229; Levine, op. cit., 1972.

102. Work of this kind, however, is going on. An important step in this direction is reported in Carrington, op. cit., 1981. Carrington shows that there is a strong positive relationship between the density of directorship interlocking among enterprises—acting as supernodes—and the degree of concentration within market-areas. Also, see Berkowitz, op. cit., forthcoming.

103. As I noted in notes 99 and 102, above, however, Carrington has shown that a relationship exists between structural relations among enterprises and the behavior of markets.

104. See Scott A. Boorman, "A Combinatorial Optimization Model for Transmission of Job Information through Contact Networks," *Bell Journal of Economics* 6 (1975): 216–49.

105. Berkowitz, op. cit., 1975; Daniel Bertaux, *Destins Personnels et Structure de Classe* (Paris: Presses Universitaires de France, 1977); Maurice Zeitlin, Lynda Ann Ewen, and Richard Earl Ratcliff, " 'New Princes' for Old? The Large Corporation and the Capitalist Class in Chile," *American Journal of Sociology* 80 (1975): 87–123.

106. Berkowitz, op. cit., 1975, pp. 132–211.

107. See Scott, op. cit., 1979, pp. 123–33.

108. Berkowitz, op. cit., 1975, pp. 213–22.

109. See Scott, op. cit., 1979.

110. Both authors largely use graphs simply to depict structures.

111. Bertaux, as translated in Scott, op. cit., 1979, p. 120.

112. This is not to argue that families themselves have become a less important component of upper-class structures but simply that they are less central to the mobilization of capital. Families remain of paramount importance, of course, in the intergenerational transfer of wealth through inheritance and in the establishment of a pool of potential recruits to fill key institutional slots. See Lorne Tepperman, "The Persistence of Dynasties: A Structural Perspective," in S. D. Berkowitz and Barry Wellman (eds.), *Structural Sociology* (Cambridge and New York: Cambridge University Press, forthcoming).

113. The argument being made here is *not* the notion of the "emergent properties" of structure which we referred to earlier, but the subtler idea that the topologies within which structural relations take place are not regular and hence that *analysts* discover new facets of these structures when they are examined sociologistically.

4: Community-Elite Networks and Markets

1. A reader who is familiar with Edward O. Laumann and Franz Pappi's *Networks of Collective Action: A Perspective on Community Influence Systems* (New York: Academic Press, 1976) will probably be struck by parallels between their interpretation of the role of experimental animals in scientific discovery and that offered here. In particular, he or she will, no doubt, find it odd that both they and I draw an analogy between sources of social scientific models and *Drosophila*. These similarities, however, are more than mere coincidence: from the beginning, structuralists have been concerned with the problem of finding ways of integrating a set of rather diverse research interests, and one way of doing this is to focus on the same research sites. Nicholas Mullins—the structuralists' resident expert on the sociology of science—first raised this issue at a conference in Camden, Maine, almost a decade ago. It has been repeated at virtually every conference or meeting in which structuralists have participated since then. So, to borrow the German phrase, this analogy has been "in the air." In a more profound sense, however, the view of science put forward here was shaped by my participation in discussions and graduate seminars at the University of Michigan and, in particular, by the ideas advanced by Professor Laumann and some of his colleagues at the time. Thus, once again, what appears on the surface to be coincidence is, in the end, a product of networks of contact!

2. It could be argued that no strict comparison can be drawn between physical artifacts or devices and the types of groups which social scientists use in their research. However, this line of reasoning would ignore the clear role of human agency in both cases: neither physical nor social artifacts are, by themselves, intrinsically usable as scientific tools. Both must be interpreted in this way by a scientist who has created a context in which they have taken on this set of meanings.

3. Strict comparisons are, of course, difficult to make because structuralists and nonstructuralists frequently abstract out different facets of these milieux for study. The strongest case can be made where, as in Granovetter's study (Mark Granovetter, *Getting a Job* [Cambridge, Mass.: Harvard University Press, 1974]), one set of events (e.g., ties among job seekers) can be analyzed and used to interpret another (e.g., rates of "success" in finding jobs).

4. This class of questions, which has to do with the general relationship between *morphology* and *behavior*, is described in Chapter 5. I argue that structural analysis will begin to have its greatest impact on the conduct of social science research when it becomes possible (a) to specify, formally, the set of constraints implicit in the patterned organization of a social system and (b) then to be able to show how these constraints shape the behavior of elements. Both of the "experimental animals" we deal with in this chapter are ideal for examining these kinds of theoretical problems because, in both cases, (a) social scientists have developed technically sophisticated tools for exploring the consequences of different behavioral models, and (b) we have at least some indication as to which morphological dimensions are most likely to be useful in modeling systemic constraints.

5. Both pluralists and antipluralists agree that groups of persons in advanced capitalist societies actively pursue their own special or particular interests. Antipluralists contend that these groups are fundamentally antidemocratic in that they seek to control governmental decision making and community issues through secret or covert means. Pluralists contend that, since there is a multiplicity of such groups, no single group predominates and that the *contention* for power or influence among them is a basic component of the democratic process. Arnold Rose, for instance, argues that:

> The relationship between the economic elite and the political authorities has been a constantly varying one of strong influence, co-operation, division of labor, and conflict, with each influencing the other in changing proportion to some extent and each operating independently of the other to a large extent. Today there is significant political control and limitation of certain activities over the economic elite, and there are also some significant processes by which the economic elite uses its wealth to help elect some political candidates and to influence other political authorities in ways which are not available to the average citizen. Further, neither the economic elite nor the political authorities are monolithic units which act with internal consensus and coordinated action with regard to each other (or probably in any other way).

(Arnold Rose, *The Power Structure: Political Process in American Society* [New York and London: Oxford University Press, 1967]).

6. This component of the pluralist argument is stated most clearly and succinctly in V. O. Key's *Politics, Parties and Pressure Groups* (New York: Knopf, 1959).

7. These are the central contentions in C. Wright Mills's classic book, *The Power*

Elite (New York: Oxford, 1956). Also see S. D. Berkowitz, "The Dynamics of Elite Structure" (Ph.D. dissertation, Brandeis University, Waltham, Mass., 1975) for a critique of Mills's work and that of his followers.

8. In many of the specifically Marxist antipluralist writings, this is referred to as the "splits in the ruling class" hypothesis. A version of this hypothesis, which became current in the U.S. in the 1960s, is presented in Carl Oglesby, *The Yankee and Cowboy War* (New York: Berkley, 1977).

9. See Delbert Miller, "Democracy and Decision-Making in the Community Power Structure," in William V. D'Antonio and Howard J. Ehrlich (eds.), *Power and Democracy in America* (Notre Dame: Notre Dame University Press, 1961), for a clear presentation of this approach.

10. The archetypical study of this kind is Floyd Hunter's *Community Power Structure: A Study of Decision Makers* (Garden City, New York: Doubleday, 1953). Robert E. Agger, "Power Attributions in the Local Community," *Social Forces* 34 (1956): 322–31; William V. D'Antonio and William H. Form, *Influentials in Two Border Cities* (Notre Dame: Notre Dame University Press, 1965).

11. John Walton, "Substance and Artifact: The Current Status of Research on Community Power Structure," *American Journal of Sociology* 71 (1966): 430–38, provides an excellent review of the relationship between the methods used in community-power research and the results obtained by analysts.

12. Robert Perrucci and Marc Pilisuk, "Leaders and Ruling Elites: The Interorganizational Bases of Community Power," *American Sociological Review* 35 (1970): 1040–57.

13. M. Herbert Danzger, "Community Power Structure: Problems and Continuities," *American Sociological Review* 29 (1964): 707–17.

14. Walton, op. cit., 1966.

15. Linton C. Freeman, Thomas J. Fararo, Warner Bloomberg, Jr., and Morris H. Sunshine, "Locating Leaders in Local Communities: A Comparison of Some Alternative Approaches," *American Sociological Review* 28 (1963): 791–98.

16. Freeman et al., op. cit., 1963; Linton C. Freeman, Warner Bloomberg, Jr., Stephen P. Koff, Morris H. Sunshine, and Thomas J. Fararo, *Local Community Leadership* (Syracuse, N.Y.: University College of Syracuse University, 1960); Linton C. Freeman, Thomas J. Fararo, Warner Bloomberg, Jr., and Morris H. Sunshine, *Metropolitan Decision-Making* (Syracuse, N.Y.: University College of Syracuse University, 1962).

17. Linton C. Freeman, *Patterns of Local Community Leadership* (Indianapolis, Ind.: Bobbs-Merrill, 1968).

18. Perrucci and Pilisuk, op. cit., 1970.

19. "Altneustadt" is a pseudonym.

20. See Edward O. Laumann and Franz Pappi, "New Directions in the Study of Community Elites," *American Sociological Review* 38 (1973): 212–30; Laumann and Pappi, op. cit., 1976.

21. Edward O. Laumann and Peter V. Marsden, "The Analysis of Oppositional Structures in Political Elites: Identifying Collective Actors," *Sozialwis-*

senschaftliche Schriften 14 (1978): 127–80. This paper was read at the International Conference on Mathematical Approaches to the Study of Social Power in Bad Homburg, West Germany, in March 1978 and reproduced in the conference volume cited above. A later version was published in the *American Sociological Review* 44 (1979): 713–32. Reference here is to the Bad Homburg version.

22. Laumann and Marsden, op. cit., 1978, p. 134. The paper examines the oppositional structure in "Towertown," a pseudonymous U.S. city, as well as "Altneustadt." Only the "Altneustadt" data are discussed here.

23. The term *color* refers to an attribute or set of attributes of the nodes in a network which may be used to discriminate between them, but which is *not* directly associated with some structural property, i.e., it is a *behavioral* attribute of the nodes. One plausible interpretation of the central task of structural analysts at the moment is thus to explain "color" from "structure."

24. Laumann and Marsden, op. cit., 1978, p. 135.

25. Ibid., p. 138.

26. Ibid., pp. 138–49.

27. Ibid., pp. 152–54.

28. Ibid. Laumann and Marsden suggest that this model corresponds to Hunter's classic depiction of power in Atlanta. See Hunter, op. cit., 1953.

29. Laumann and Marsden, op. cit., 1978, pp. 154–55.

30. Ibid., p. 156.

31. Ibid., p. 157.

32. Their third model, of course, is also a sort of pluralist one, but there is no crosscutting structure of alignments which I take to represent the classic pluralist position. See Key, op. cit., 1959.

33. Laumann and Marsden, op. cit., 1978, p. 162. The smallest space model presented in the version of this paper which appeared in the *American Sociological Review* is slightly different because of the use of a different proximity measure. I refer to the Bad Homburg paper here for simplicity's sake.

34. Harrison C. White, Scott A. Boorman, and Ronald L. Breiger, "Social Structure from Multiple Networks, I: Blockmodels of Roles and Positions," *American Journal of Sociology* 81 (1976): 730–80; Phipps Arabie, Scott A. Boorman, and Paul Levitt, "Constructing Blockmodels: How and Why," *Journal of Mathematical Psychology* 17 (1978): 21–63.

35. This description—and several others which follow in this section—is condensed from Peter J. Carrington, Greg H. Heil, and Stephen D. Berkowitz, "A Goodness-of-Fit Index for Blockmodels," *Social Networks* 2 (1979/80): 219–34.

36. In practice, three kinds of criteria are used to "fit" image graphs to datagraphs. "Fat fit" requires that *only* 1s may be mapped into one-blocks and 0s into zero-blocks. "Lean-fit" requires that zero-blocks be "pure," i.e.,

that they contain only 0s. "α fit" requires that a cut-off value, α, between 0 and 1 be specified and that there be 1s in image matrices wherever data blocks have densities of 1s greater than this value, and 0s wherever they have less than it (Arabie, Boorman, and Levitt, op. cit., 1978). Since the requirements for "fat fit" are often too strong, i.e., they correspond to those in the Lorrain-White model, and those for "lean fit" are both too strong (in the case of zero-blocks) and too weak (in the case of one-blocks), variations on the α fit criterion are now most commonly used. Carrington, Heil, and Berkowitz, op. cit., 1979/80, offer a method for interpreting the goodness-of-fit of image graphs to datagraphs.

37. In strict terms, it is not necessary in blockmodeling to prohibit ties. As we note in note 36, above, α fit criteria will tolerate "impure" blocks, i.e., ones which contain nonspecified ties. In this case, because our discussion refers back to an earlier example which employs the Lorrain-White method, it is necessary to include these prohibitions in order to make the two methods comparable. For a further discussion of this point, see Greg H. Heil and Harrison C. White, "An Algorithm for Finding Simultaneous Homomorphic Correspondences between Graphs and Their Image Graphs," *Behavioral Science* 21 (1976): 26–35.

38. Internal ties are those between members of the same bloc, i.e., a subset of nodes clustered on the basis of structural similarity. External ties are between blocs.

39. As in the Lorrain-White method, blockmodels can be falsified: when there is poor fit between a hypothesized structure and the data themselves, then we are forced to reject the interpretation or hypotheses embodied in a given "image" graph. Similarly, when a given model-image is highly permissive, we are immediately aware of this and therefore less likely to draw strong conclusions from a "good fit" between weak hypotheses and the data.

40. Ronald L. Breiger, "Toward an Operational Theory of Community Elite Structure," *Quality and Quantity* 13 (1979): 21–47. See James S. Coleman, *Community Conflict* (New York: Free Press, 1957); Peter H. Rossi, "Power and Community Structure," *Midwest Journal of Political Science* 4 (1960): 390–401; Anthony Sampson, *Anatomy of Britain* (London: Holder and Stoughton, 1962); Hunter, op. cit., 1953; Robert S. and Helen M. Lynd, *Middletown* (New York: Harcourt, Brace, 1929); Robert A. Dahl, *Who Governs?* (New Haven: Yale University Press, 1961).

41. Breiger, op. cit., 1979, p. 26.

42. Ibid.

43. Ibid., pp. 27–29.

44. Ibid., pp. 30–31.

45. Ibid., pp. 34–39.

46. See note 36, above.

47. Breiger, op. cit., 1979, pp. 39–45.

48. Here we can see one of the principal strengths of blockmodeling: it is "fine-grained," i.e., it allows researchers to consider subtle patterning in the relationships among elements. Precisely because of this, however, it is often

a poor technique to use where a researcher has only a rudimentary idea of the patterns he or she may encounter. In this sense, it is *too* precise. The obvious way out of this dilemma is the one suggested here: to employ more broadly gauged techniques in the exploratory phases of a piece of research and, only then, to begin to formulate and test blockmodels.

49. Thomas Kuhn, *The Structure of Scientific Revolutions* (Chicago: University of Chicago Press, 1970), pp. 52–65.

50. See Abraham Kaplan, *The Conduct of Inquiry: Methodology for Behavioral Science* (Scranton, Pa.: Chandler, 1964), pp. 55–56.

51. S. D. Berkowitz, "Markets and Market-Areas," in S. D. Berkowitz and Barry Wellman (eds.), *Structural Sociology* (Cambridge and New York: Cambridge University Press, forthcoming), begins with the conventional buyer and seller roles, but extends these on the basis of a more explicit notion of the nature of transactions. This argument is too extensive to be reproduced here, but consider, for example, the well-recognized role of the "middle-man" who is, simultaneously, a buyer and seller or "buyer-seller."

52. The use of the term *market* without concrete referents is now commonplace in economics. For instance, Milton Friedman argues that:

> The basic inequality, it should be noted, is inequality in the ownership of resources. What the *market* does is primarily to determine the return per unit resource, and there is no reason to believe that the *market* aggravates the inequality in the ownership of resources. Moreover, any given degree of inequality is much more serious in an economy that is governed largely by status or tradition than in a *market* economy where there is much chance for shifts in the ownership of resources. Historically, the fundamental inequality of economic status has been and is almost certainly greater in economies that do not rely on the free *market* than in those that do. [Milton Friedman, *Price Theory* (Chicago: Aldine, 1976), p. 11; emphasis mine]

Note that in the first instance the reference is to a *process* for determining the price and allocation of goods or services. Friedman has no *particular* market in mind, nor, in fact, could he generalize as he does if he did. The second use is more specific. Here the market is at least a *construct*, i.e., a concept which cannot be operationalized directly, but which can be detected indirectly through others. In this second case, we could *measure* the extent to which a market existed by reference to supply and demand and the consequent distribution of income. The third use of the term in this paragraph refers to a model of an *economy*—or a point on a continuum of such models—which is only meaningful in contrast to something else, i.e., a "nonmarket" economy. In this case, neither term is clearly defined nor associated with some set of observables. Finally, we have the term *free market*, which most economists avoid because it immediately confuses ideal-typical models and "reality." No one has produced convincing evidence that any such thing ever existed and, in fact, use of the term in this way reflects a lack of appreciation for the differences between models and the things they seek to represent. Similarly vague or slippery uses of the term occur in many standard texts. See Everett E. Hagen, *The Economics of Development* (Homewood, Ill.: Irwin, 1968), pp. 406–7.

53. Edward H. Chamberlain, *The Theory of Monopolistic Competition* (Cambridge, Mass.: Harvard University Press, 1933).

54. A. Michael Spence, *Market Signaling: Informational Transfer in Hiring and Related Screening Processes* (Cambridge, Mass.: Harvard University Press, 1974).

55. See John C. Harsanyi, *Rational Behavior and Bargaining Equilibrium in Games and Social Situations* (New York and Cambridge: Cambridge University Press, 1977).

56. Nancy Howell, *The Search for an Abortionist* (Chicago: University of Chicago Press, 1969).

57. Granovetter, op. cit., 1974.

58. Ibid., p. 38.

59. Ibid., pp. 30–35.

60. Ibid., pp. 37–39.

61. Scott A. Boorman, "A Combinatorial Optimization Model for Transmission of Job Information through Contact Networks," *Bell Journal of Economics* 6 (1975): 216–49.

62. See John Von Neumann and Oskar Morgenstern, *Theory of Games and Economic Behavior* (New York: Wiley, 1944).

63. Tight formal models of this kind have the disadvantage that they are relatively easy to disprove given that even small variations in one or two parameters will tend to throw them off. Under these circumstances, Boorman's formalization needed to be far more accurate than it would have had to be if he had not presented it in this way.

64. See Harriet Friedmann, "The Transformation of Wheat Production in the Era of the World Market, 1873–1935: A Global Analysis of Production and Exchange" (Ph.D. dissertation, Harvard University, Cambridge, Mass., 1976); Harriet Friedmann, "Simple Commodity Production and Wage Labour in the American Plains," *Journal of Peasant Studies* 6 (1978a): 71–100.

65. This aspect of her work comes through most clearly in "World Market, State, and Family Farm: Social Bases of Household Production in the Era of Wage Labor," *Comparative Studies in Society and History* 20 (1978b): 545–86. Note here, also, the essential features of a theoretically complete, though informal, structuralist model: (a) a developed set of constraints which (b) set parameters for actors, eventuating in (c) a structured or patterned set of relations among them.

66. Harrison C. White, "Subcontracting with an Oligopoly: Spence Revisited," *RIAS Working Paper #1* (Cambridge, Mass.: Harvard University, 1976); Harrison C. White, "On Markets," *RIAS Program Working Paper #16* (Cambridge, Mass.: Harvard University, 1979); Harrison C. White, "Markets as Stages for Producers" (Toronto: University of Toronto, Structural Analysis Programme, 1980). Also see Harrison C. White, "Production Markets as Induced Role Structures," in Samuel Leinhardt (ed.), *Sociological Methodology, 1981* (San Francisco: Jossey-Bass, 1981); Harrison C. White, "Where Do Markets Come From?" *American Journal of Sociology* (in press); Harrison C. White, "Varieties of Markets," in S. D. Berkowitz and Barry Wellman (eds.), *Structural Sociology* (Cambridge and New York: Cambridge University Press, forthcoming).

5: The Future of Structural Analysis

1. References to many works that do not fit easily into the present narrative have been included in the listings of "additional sources" at the end of each chapter, or in the bibliography. Readers will note that a number of important articles and books in mathematical sociology have not been cited here. For reasons that should be clear by the end of this chapter, I thought they were less central to structural analysis, per se, and space did not permit discussion of all of them. Other definitions of the boundaries of the field are, of course, possible. In some instances, I have avoided detailed discussion of recent work in "old" branches of the field. Jeremy F. Boissevain and J. Clyde Mitchell (*Network Analysis: Studies in Human Interaction* [Mouton: The Hague, 1973]), for instance, have compiled an excellent reader that incorporates work that continues the tradition of the "first wave" social anthropologists. These, and similar developments, will have to await later treatment.

2. Given the design of this book, many of the more technically sophisticated developments in the area have not been discussed in detail. In fact, many of the more technical aspects of work described here have been omitted. I have tried, where practical, at least to refer to this literature. Readers with well-established mathematical skills are referred to the bibliographies cited in note 3, below, for more intensive coverage of these aspects of the field.

3. Ronald S. Burt, "Models of Network Structure," *Annual Review of Sociology* 6 (1980): 79–141; Linton C. Freeman, *A Bibliography of Social Networks* (Monticello, Ill.: Council of Planning Librarians, 1976); Samuel Leinhardt (ed.), *Social Networks: A Developing Paradigm* (New York: Academic Press, 1977); and Barry Wellman, "Network Analysis: From Metaphor and Method to Theory and Substance," *Sociological Theory* 1, forthcoming.

4. More than most paradigms in sociology—indeed, in the social sciences generally—structural analysis seems to have developed in response to a number of rather difficult technical problems that grew out of a few clearly defined and restricted theoretical issues, e.g., what is the relationship between *morphology* and *behavior*? In this sense, its history is more like that of, say, quantum physics than symbolic interactionism or critical theory. "Theory" and "methods" have been heavily intertwined and it makes little sense to teach one without reference to the other. Similarly, the "substantive" questions to which structural analysts have made the greatest contribution have all been ones where it is impossible to avoid fundamental theoretical and methodological issues. So, in the same fashion that it would be a sign of impending intellectual decay for one to suggest teaching a course on the "results" of experiments in physical science without placing them in a theoretical and methodological context, it is impossible to separate out the various components of research practice in structural analysis.

5. Thomas Kuhn, *The Structure of Scientific Revolutions* (Chicago: University of Chicago Press, 1970), pp. 10–22.

6. Ibid., p. 11.

7. Ibid., pp. 12–18.

8. Ibid., pp. 18–22.

9. "Straightforward" not in the sense that they are simple or trivial, but that the problems involved are "closed." See Kuhn, op. cit., 1970, pp. 19–20.

10. Kuhn, op. cit., 1970, pp. 36–39, 45–49, 52–61.

11. Ibid., pp. 62–76.

12. Ibid., pp. 92–110.

13. See Nicholas Mullins, *Theory and Theory Groups in Contemporary American Sociology* (New York: Harper & Row, 1973), for a summary of the principal interpretative schools and paradigms in sociology. One could argue, given the non-physical-science-like development of the discipline, that there are *no* paradigms in sociology, since no one framework predominates. I think this is semantic quibbling: the particular framework of each set of practitioners behaves in much the same way that paradigms do in Kuhn's classic model. And some "paradigms" are, in fact, "replaced" by others in the same way. Given what might be more appropriately referred to as multiple "branches" of the discipline, there are clear chains of succession. This approach would suggest that sociology is not a "multiparadigm discipline" but a set of disciplines masquerading as one.

14. A *scientific community*, in this sense, consists of the body of practitioners actively pursuing a common set of problems within the context of a shared set of interpretations of their tasks (Kuhn, op. cit., 1970, pp. 18–22). I do not intend the more global sense of the term, i.e., "men and women of science." Kuhn uses the term *scientific group* (ibid.). I prefer the term I use here because it conveys more of a sense of shared expectations.

15. Pre-eminently, Harvard, Michigan, and Chicago.

16. Mullins, op. cit., 1973.

17. Mullins, op. cit., 1973, defines it as such. As far as I know, he was the first to use the term consistently to refer to the range of activities and perspectives now incorporated within the area. S. F. Nadel (*The Theory of Social Structure* [Glencoe, Ill.: Free Press, 1957]) defines a task he calls "a structural analysis" which is probably the point of origin of the present term. For many years, the term "network analysis" was common. As I noted earlier, this usage confuses the object of a field of study with its most common tool, i.e., "social networks."

18. These indices of the establishment of a paradigm follow from Kuhn, op. cit., 1970, p. 19.

19. See A. J. Coale, *The Growth and Structure of Human Populations: A Mathematical Investigation* (Princeton: Princeton University Press, 1972); J. A. Menken, "Biological Determinants of Demographic Processes," *American Journal of Public Health* 64 (1974): 657–61.

20. Nancy Howell, *Demography of the Dobe !Kung* (New York: Academic Press, 1979).

21. For a description of the techniques used in doing this, see Nancy Howell and Victor Lehotay, "AMBUSH: A Computer Program for Stochastic Microsimulation of Small Human Populations," *American Anthropologist* 80 (1978): 905–22.

22. Peter J. Carrington, "Horizontal Cooptation through Corporate Interlocks" (Ph.D. dissertation, University of Toronto, 1981); Scott A. Boorman and Paul R. Levitt, *The Genetics of Altruism* (New York: Academic Press, 1980).

23. Kuhn, op. cit., 1970, p. viii.

24. Laumann and his colleagues, in the works cited throughout this volume, have probably made the most consistent effort to link research practice to a well-articulated and general body of theory. To note that this has been the exception rather than the rule is not to say that researchers have not been making these connections, nor that a systematic body of structural analytic theory does not exist. Rather, it is simply to argue that the *systemic* aspects of this framework have not been presented in a thoroughgoing fashion.

25. I use the term *conditions* here to mean that each dimension dynamically interacts with and shapes the development of the others. I avoid the term *determines* here because it suggests linear causality. See Karl Marx, *A Contribution to the Critique of Political Economy* (Moscow: Progress Publishers, n.d.), and François Lorrain, *Réseaux Sociaux et Classifications Sociales* (Paris: Hermann, 1975), for similar uses of the concept.

26. The "scope" of a structure is the specific set or sets of elements that it includes. The "functions" of a structure are the recognized ways in which that structure, as a whole, interacts with its environment. Its "role organization" is the abstract division of labor among its parts.

27. An *embedding dimension* is, in practice, the concrete, measurable activity upon which an analyst chooses to focus. Thus, if one were interested in the intergenerational transmission of "wealth" and "social class," one might choose "kinship" as his or her embedding dimension. See S. D. Berkowitz, "The Dynamics of Elite Structure: A Critique of C. Wright Mills' 'Power Elite' Model" (Ph.D. dissertation, Brandeis University, Waltham, Mass., 1975).

28. See the exercises for Chapter 2 for a concrete illustration of this problem.

29. If, by contrast, *different* mapping rules are followed, such comparisons are meaningless. For instance, if we are mapping two or more kinship structures and in one we permit ties between siblings (i.e., a "tie" indicates "is related to"), and in the other we do not, a distance measure between nodes (e.g., "*A* and *B* are at t^3 from one another") will mean different things in the two models.

30. See Anatol Rapoport, *Fights, Games, and Debates* (Ann Arbor: University of Michigan Press, 1974); Anatol Rapoport, *Two Person Game Theory: The Essential Ideas* (Ann Arbor: University of Michigan Press, 1973); Anatol Rapoport, *N-Person Game Theory: Concepts and Applications* (Ann Arbor: University of Michigan Press, 1970). Also see, R. D. Luce and A. W. Tucker (eds.), *Contributions to the Theory of Games, IV* (Princeton: Princeton University Press, 1959).

31. Edward O. Laumann and Peter V. Marsden, "The Analysis of Oppositional Structures in Political Elites: Identifying Collective Actors," *American Sociological Review* 44 (1979): 713–32.

32. Ronald L. Breiger, "Toward an Operational Theory of Community Elite Structure," *Quality and Quantity* 13 (1979): 21–47; Scott A. Boorman, "A Combinatorial Optimization Model for Transmission of Job Information through Contact Networks," *Bell Journal of Economics* 6 (1975): 216–49.

33. Ronald S. Burt, "A Structural Theory of Interlocking Corporate Directorates," *Social Networks* 1 (1978/79): 415–35.

34. James S. Coleman, *The Adolescent Society: The Social Life of the Teenager and Its Impact on Education* (New York: Free Press, 1961).

35. See Charles Tilly, *From Mobilization to Revolution* (Reading, Mass.: Addison-Wesley, 1978); Charles Tilly, "Food Supply and Public Order in Modern Europe," in Charles Tilly (ed.), *The Formation of National States in Western Europe* (Princeton: Princeton University Press, 1975).

36. This is particularly evident in studies of (a) interlocking directorates and corporate structure, (b) markets and market structure, (c) search processes of various kinds, and (d) microsimulations.

37. See Douglas White, "Material Entailment Analysis: Theory and Illustrations" (Research Report 15, School of Social Sciences, University of Calif., Irvine, 1980).

38. Howell and Lehotay, op. cit., 1978; John L. Delaney, "Network Dynamics for the Weak Tie Problem: A Simulation Study" (Cambridge, Mass.: Harvard-Yale Preprints in Mathematical Sociology, 1978).

39. In addition to the work cited earlier, see Phillip Bonacich, "Using Boolean Algebra to Analyze Overlapping Memberships," in K. F. Schuessler (ed.), *Sociological Methodology 1978* (San Francisco: Jossey-Bass, 1978); Phillip Bonacich, "The Common Structure Semigroup: An Alternative to the Boorman and White 'Joint Reduction,'" *American Journal of Sociology*, forthcoming; Scott A. Boorman and Harrison C. White, "Social Structure from Multiple Networks, II: Role Structures," *American Journal of Sociology* 81 (1976): 1384–446; Peter J. Carrington and Greg H. Heil, "A Method for Finding Blockmodels of Networks," *Journal of Mathematical Sociology*, forthcoming.

40. Peter J. Carrington, Greg H. Heil, and Stephen D. Berkowitz, "A Goodness-of-Fit Index for Blockmodels," *Social Networks* 2 (1979/80): 219–34.

41. What is needed here is some way of assessing, jointly, a whole set of probability distributions, i.e., the likelihood of getting a given number of ones and zeros in each block within a matrix.

42. The b-measure calculates the extent to which the densities of 1s and 0s in the submatrices of the blocked data matrices vary along a continuum from "purity" to the worst tolerable density (α). The index has its lowest value (0) in the case of worst possible fit and increases monotonically as block densities approach 0 for zeroblocks and 1 for oneblocks. Its maximum value, therefore, is 1 at purity. See Carrington, Heil, and Berkowitz, op. cit., 1979/80, pp. 226–31.

43. R. H. Atkin, *Combinatorial Connectivities in Social Systems* (Basel: Birkhauser, 1977); R. H. Atkin, "An Algebra for Patterns on a Complex, I," *International Journal of Man-Machine Studies* 6 (1974): 285–307; R. H. Atkin, "An Algebra for Patterns on a Complex, II," *International Journal of Man-Machine Studies* 8 (1976): 483–88.

44. For an excellent discussion of the potential role of q-analysis in structuralist research—and a number of related issues—see Patrick Doreian, "On the Evolution of Group and Network Structure," *Social Networks* 2 (1979/80): 235–52.

45. See Ludwig von Bertalanffy, "The History and Status of General Systems Theory," in George J. Klir (ed.), *Trends in General Systems Theory* (New York: Wiley, 1972), pp. 21–41.

46. The essential idea behind general systems theory, of course, was and is that practitioners from each of several disciplines would translate their problems and findings into a sort of common language such that everyone could draw upon and synthesize a set of pan-scientific principles and observations (Ludwig von Bertalanffy, *General System Theory* [New York: George Braziller, 1968]). The difficulty with this approach is that it often (a) leads practitioners to formulate extremely general statements in order to increase compatibilities between their findings or (b) simply becomes a way of saying the same things in a different set of words. Thus, it tends to encourage unnecessary generality or the presentation of prosaic ideas in new garb. The best work in the area, as a result, seems to be either (a) quite mathematical—that is to say, abstract and precise, but without immediate applications to scientific problems or (b) based almost exclusively in one discipline, but attempting to translate the problems of *another* field or area into its terms.

47. I argue that the most important consequence of focusing a set of researchers on a few carefully chosen experimental animals is that these "animals" themselves tend to act as a brake on the natural tendency of scientists to become abstracted from the "real world" they are trying to describe and interpret. At the same time, because they are examining a common research site, scientists are forced to take one another's perspectives and findings into account. Neither of these conditions were and are typically met in much of the interdisciplinary or transdisciplinary work in general systems research.

48. Ferdinand de Saussure, *Course in General Linguistics* (New York: McGraw-Hill, 1966); Roman Jakobson, *Selected Writings*, 4 vols. (The Hague: Mouton, 1962); Roman Jakobson and Morris Halle, *Fundamentals of Language* (The Hague: Mouton, 1956); Edward Sapir, *Language* (New York: Harcourt, Brace, 1921); Jean Piaget, *Structuralism* (London: Routledge & Kegan Paul, 1971).

49. See Terence Hawkes, *Structuralism and Semiotics* (Berkeley and Los Angeles, Calif.: University of California Press, 1977), for a discussion of interchanges between structural psychology, linguistics, and anthropology. Also see Lorrain, op. cit., 1975.

50. This position, of course, tends toward absolute relativism at the extreme, i.e., classic nineteenth-century idealism. I argue that this is inconsistent with scientific reasoning in general.

51. See Claude Lévi-Strauss, *The Elementary Structures of Kinship* (Boston: Beacon Press, 1969), for suggestions of this. Also see Noam Chomsky, "Formal Properties of Grammars," in R. D. Luce, R. R. Bush, and E. Galanter (eds.), *Handbook of Mathematical Psychology*, vol. 2 (New York: Wiley, 1963).

Bibliography

Aaronvitch, Sam. *Monopoly: A Study of British Monopoly Capitalism*. London: Lawrence & Wishart, 1955.

———. *The Ruling Class: A Study of British Finance Capital*. London: Lawrence & Wishart, 1961.

Abell, P. "Measurement in Sociology: I. Measurement and Systems." *Sociology* 2 (1968): 1–80.

———. "Measurement in Sociology: II. Measurement Structure and Sociological Theory." *Sociology* 3 (1969): 397–411.

———. *Model Building in Sociology*. London: Weidenfeld and Nicolson, 1971.

Agger, Robert E. "Power Attributions in the Local Community." *Social Forces* 34 (1956): 322–31.

Alba, R. "A Graph-Theoretic Definition of a Sociometric Clique." *Journal of Mathematical Sociology* 3 (1973): 113–26.

Alba, R., and C. Kadushin. "The Intersection of Social Circles: A New Measure of Social Proximity in Networks." *Sociological Methods and Research* 5 (1976): 77–103.

Alba, R., and G. Moore. "Elite Social Circles." *Sociological Methods and Research* 7 (1978): 167–88.

Alt, J., and N. Schofield. "Clique Analysis of a Tolerance Relation." *Journal of Mathematical Sociology* 6 (1978): 155–62.

Arabie, Phipps. "Clustering Representations of Group Overlap." *Journal of Mathematical Sociology* 5 (1977): 113–28.

Arabie, Phipps; Scott A. Boorman; and Paul Levitt. "Constructing Blockmodels: How and Why." *Journal of Mathematical Psychology* 17 (1978): 21–63.

Arabie, Phipps, and J. Douglas Carroll. "MAPCLUS: A Mathematical Programming Approach to Fitting the ADCLUS Model." *Psychometrika* 45 (1980): 211–35.

Arrow, K. J.; S. Karlin; and P. Suppes (eds.). *Mathematical Models in the Social Sciences*. Stanford: Stanford University Press, 1960.

Atkin, R. H. "Multi-Dimensional Structure in the Game of Chess." *International Journal of Man-Machine Studies* 4 (1972): 341–62.

———. "An Algebra for Patterns on a Complex, I." *International Journal of Man-Machine Studies* 6 (1974): 285–307.

Semicolons are used to separate the names of authors in multiauthor works, not to denote the seniority of any author.

_____. "An Algebra for Patterns on a Complex, II." *International Journal of Man-Machine Studies* 8 (1976): 483–88.

_____. *Combinatorial Connectivities in Social Systems*. Basel: Birkhauser, 1977.

Bailey, Norman T. J. *The Mathematical Theory of Infectious Diseases and Its Applications*. New York: Hafner, 1975.

Bain, Joe S. *Industrial Organization*. New York: John Wiley & Sons, 1968.

Barnes, John A. "Class and Committees in a Norwegian Island Parish." *Human Relations* 7 (1954): 39–58.

_____. "Graph Theory and Social Networks: A Technical Comment on Connectedness and Connectivity." *Sociology* 3 (1969): 215–32.

_____. *Social Networks*. Reading, Mass.: Addison-Wesley, 1972.

Barratt Brown, Michael. "The Controllers of British Industry." In K. Coates (ed.), *Ca n the Workers Run Industry?* London: Sphere, 1968.

Bearden, James; William Atwood; Peter Freitag; Carol Hendricks; Beth Mintz; and Michael Schwartz. "The Nature and Extent of Bank Centrality in Corporate Networks." Paper read at the annual meetings of the American Sociological Association, 1975.

Beauchamp, Murray. "An Improved Index of Centrality." *Behavioral Science* 10 (1965): 161–63.

Bergin, Thomas Goddard, and Max Harold Fisch (eds.). *The New Science of Giambattista Vico*. Ithaca: Cornell University Press, 1970.

Berkowitz, S. D. "The Dynamics of Elite Structure: A Critique of C. Wright Mills' 'Power Elite' Model." Ph.D. dissertation, Brandeis University, Waltham, Mass., 1975.

_____. "Structural and Non-structural Models of Elites." *Canadian Journal of Sociology* 5 (1980): 13–30.

_____. "Markets and Market-Areas." In S. D. Berkowitz and Barry Wellman (eds.), *Structural Sociology*. Cambridge and New York: Cambridge University Press, forthcoming.

Berkowitz, S. D.; P. J. Carrington; J. S. Corman; and L. Waverman. "Flexible Design for a Large-scale Corporate Data Base." *Social Networks* 2 (1979): 75–83.

Berkowitz, S. D.; P. J. Carrington; Yehuda Kotowitz; and Leonard Waverman. "The Determination of Enterprise Groupings through Combined Ownership and Directorship Ties." *Social Networks* 1 (1978/79): 391–413.

Berkowitz, S. D., and L. Felt. *Phase I of a Structural Analysis of the Canadian Financial System*. 2 vols. Report. Toronto: Institute for Policy Analysis, 1975.

Berkowitz, S. D., and Greg Heil. "Dualities in Methods of Social Network Research." Toronto: Structural Analysis Programme, 1980.

Berkowitz, S. D.; Yehuda Kotowitz; and Leonard Waverman; with Bruce Becker, Randy Bradford, Peter Carrington, June Corman, and Gregory Heil. *Enterprise Structure and Corporate Concentration*. Ottawa: Royal Commission on Corporate Concentration, 1976. (Issued 1978.)

Berkowitz, S. D., and Barry Wellman (eds.). *Structural Sociology*. Cambridge and New York: Cambridge University Press, forthcoming.

Berle, Adolph A., and Gardiner C. Means. *The Modern Corporation and Private Property*. New York: Macmillan, 1932.

Bernard, H. Russell, and Peter D. Killworth. "On the Social Structure of an Ocean-Going Research Vessel and Other Important Things." *Social Science Research* 2 (1973): 145–84.

Bernard, H. Russell; Peter D. Killworth; and Lee Sailer. "Informant Accuracy in Social Network Data IV: A Comparison of Clique-Level Structure in Behavioral and Cognitive Network Data." *Social Networks* 2 (1979/80): 191–218.

Bertalanffy, Ludwig von. *General System Theory*. New York: George Braziller, 1968.

————. "The History and Status of General Systems Theory." In George J. Klir (ed.), *Trends in General Systems Theory*. New York: Wiley, 1972.

Bertaux, Daniel. *Destins Personnels et Structure de Classe*. Paris: Presses Universitaires de France, 1977.

Black, Donald. "The Boundaries of Legal Sociology." *Yale Law Journal* 81 (1972): 1086–100.

Blau, Peter M. *Exchange and Power in Social Life*. New York: Wiley, 1964.

————. "Parameters of Social Structure." *American Sociological Review* 39 (1974): 615–35.

Boissevain, Jeremy F. *Friends of Friends*. Oxford: Basil Blackwell, 1974.

Boissevain, Jeremy F., and J. Clyde Mitchell (eds.). *Network Analysis: Studies in Human Interaction*. Mouton: The Hague, 1973.

Bonacich, Phillip. "Technique for Analyzing Overlapping Memberships." In Herbert L. Costner (ed.), *Sociological Methodology, 1972*. San Francisco: Jossey-Bass, 1972.

————. "Using Boolean Algebra to Analyze Overlapping Memberships." In K. F. Schuessler (ed.), *Sociological Methodology, 1978*. San Francisco: Jossey-Bass, 1978.

————. "The Common Structure Semigroup: An Alternative to the Boorman and White 'Joint Reduction.'" *American Journal of Sociology*, forthcoming.

Boorman, Scott A. "A Combinatorial Optimization Model for Transmission of Job Information Through Contact Networks." *Bell Journal of Economics* 6 (1975): 216–49.

Boorman, Scott A., and Phipps Arabie. "Algebraic Approaches to the Comparison of Concrete Social Structures Represented as Networks: Reply to Bonacich." *American Journal of Sociology*, forthcoming.

Boorman, Scott A., and Paul R. Levitt. *The Genetics of Altruism*. New York: Academic Press, 1980.

Boorman, Scott A., and Harrison C. White. "Social Structure from Multiple Networks, II: Role Structures." *American Journal of Sociology* 81 (1976): 1384–446.

Boswell, David. "Personal Crises and the Mobilization of the Social Network." In J. Clyde Mitchell (ed.), *Social Networks in Urban Situations: Analyses of Personal Relationships in African Towns*. Manchester: Manchester University Press, 1969.

Bott, Elizabeth. *Family and Social Network: Roles, Norms, and External Relationships in Ordinary Urban Families*. London: Tavistock, 1957.

Boyd, John Paul. "The Algebra of Kinship." Doctoral dissertation, University of Michigan, Ann Arbor, 1966.

_____. "The Algebra of Group Kinship." *Journal of Mathematical Psychology* 6 (1969): 139–67.

_____. "The Universal Semigroup of Relations." *Social Networks* 2 (1979/80): 91–117.

Boyd, John Paul; H. Haehl; and L. D. Sailer. "Kinship Systems and Inverse Semigroups." *Journal of Mathematical Sociology* 2 (1971): 37–61.

Boyd, John Paul, and William Livant. "Some Properties and Implications of Lexical Trees." Ann Arbor: University of Michigan MS, 1964.

Breiger, Ronald L. "The Duality of Persons and Groups." *Social Forces* 53 (1974): 181–90.

_____. "Toward an Operational Theory of Community Elite Structure." *Quality and Quantity* 13 (1979): 21–47.

Breiger, Ronald L.; Scott A. Boorman; and Phipps Arabie. "An Algorithm for Clustering Relational Data, with Applications to Social Network Analysis and Comparison with Multi-Dimensional Scaling." *Journal of Mathematical Psychology* 12 (1975): 328–83.

Breiger, Ronald L., and Philippa E. Pattison. "The Joint Role Structure of Two Communities' Elites." *Sociological Methods and Reseach* 7 (1978): 213–26.

Buckley, Walter. *Sociology and Modern Systems Theory*. Englewood, Cliffs, N.J.: Prentice-Hall, 1967.

Burt, Ronald S. "Positions in Multiple Network Systems, Part One: A General Conception of Stratification and Prestige in a System of Actors Cast as a Social Typology." *Social Forces* 56 (1977): 106–31.

_____. "Applied Network Analysis: An Overview." *Sociological Methods and Research* 7 (1978): 123–31.

_____. "Stratification and Prestige Among Elite Experts in Methodological and Mathematical Sociology Circa 1975." *Social Networks* 1 (1978/79): 105–58.

_____. "A Structural Theory of Interlocking Corporate Directorates." *Social Networks* 1 (1978/79): 415–35.

_____. "Disaggregating the Effect on Profits in Manufacturing Industries of Having Imperfectly Competitive Consumers and Suppliers." *Social Science Research* 8 (1979): 120–43.

_____. "Models of Network Structure." *Annual Review of Sociology* 6 (1980): 79–141.

———. "Cooptive Corporate Actor Networks: A Reconsideration of Interlocking Directorates Involving American Manufacturing." *Administrative Science Quarterly*, forthcoming.

Busacker, Robert G., and Thomas L. Saaty. *Finite Graphs and Networks: An Introduction with Applications*. New York: McGraw-Hill, 1965.

Capobianco, Michael. "Statistical Inference in Finite Populations Having Structure." *Transactions of the New York Academy of Sciences* 32 (1970): 401–13.

Capobianco, Michael, and John C. Molluzzo. "The Strength of a Graph and its Application to Organizational Structure." *Social Networks* 2 (1979/80): 275–83.

Carrington, Peter J. "Horizontal Cooptation through Corporate Interlocks." Ph.D. dissertation, University of Toronto, 1981.

Carrington, Peter J., and Greg H. Heil. "A Method for Finding Blockmodels of Networks." *Journal of Mathematical Sociology*, forthcoming.

Carrington, Peter J.; Greg H. Heil; and Stephen D. Berkowitz. "A Goodness-of-Fit Index for Blockmodels." *Social Networks* 2 (1979/80): 219–34.

Carroll, William; John Fox; and Michael Ornstein. "The Network of Directorate Interlocks among the Largest Canadian Firms." Downsview, Ontario: Institute for Behavioral Research, York University, 1977.

Cartwright, Dorwin, and Frank Harary. "Structural Balance: A Generalization of Heider's Theory." *Psychological Review* 63 (1956): 277–93.

Caves, Richard. *American Industry: Structure, Conduct, Performance*. 4th ed. Englewood Cliffs, N.J.: Prentice-Hall, 1977.

Chamberlain, Edward H. *The Theory of Monopolistic Competition*. Cambridge, Mass.: Harvard University Press, 1933.

Chevalier, Jean-Marie. *La Structure Financière de l'Industrie Américaine*. Paris: Cujas, 1970.

Chomsky, Noam. "Formal Properties of Grammars." In R. D. Luce, R. R. Bush, and E. Galanter (eds.), *Handbook of Mathematical Psychology*. Vol. 2. New York: Wiley, 1963.

Christensen, Harold (ed.). *Handbook of Marriage and the Family*. Chicago: Rand McNally, 1964.

Coale, A. J. *The Growth and Structure of Human Populations: A Mathematical Investigation*. Princeton: Princeton University Press, 1972.

Cohen, Abner. *Custom and Politics in Urban Africa*. Berkeley: University of California Press, 1969.

Coleman, James S. *Community Conflict*. New York: Free Press, 1957.

———. *The Adolescent Society: The Social Life of the Teenager and Its Impact on Education*. New York: Free Press, 1961.

———. *An Introduction to Mathematical Sociology*. New York: Free Press, 1964.

———. "Clustering in N Dimensions by Use of a System of Forces." *Journal of Mathematical Sociology* 1 (1970): 1–47.

_____. *The Mathematics of Collective Action*. Chicago: Aldine, 1973.

Coleman, James S.; Elihu Katz; and Herbert Menzel. *Medical Innovation: A Diffusion Study*. Indianapolis: Bobbs-Merrill, 1966.

Coleman, James S., and D. MacRae. "Electronic Data Processing of Sociometric Data for Groups up to 1000 in Size." *American Sociological Review* 25 (1960): 722–26.

Coombs, Clyde H. *A Theory of Data*. New York: Wiley, 1964.

Coombs, G. "Networks and Exchange: The Role of Social Relationships in a Small Voluntary Association." *Journal of Anthropological Research* 29 (1973): 96–112.

_____. "Opportunities, Information Networks, and the Migration-Distance Relationship." *Social Networks* 1 (1978/79): 257–76.

Costner, Herbert L. *Sociological Methodology*. San Francisco: Jossey-Bass, 1972.

Coxon, A. P. M., and Charles Jones. "Multidimensional Scaling: Exploration to Confirmation." *Quality and Quantity* 14 (1980): 31–73.

Craven, Paul, and Barry Wellman. "The Network City." *Sociological Inquiry* 43 (1973): 57–88.

Dahl, Robert A. *Who Governs?* New Haven: Yale University Press, 1961.

D'Antonio, William V., and Howard J. Ehrlich (eds.). *Power and Democracy in America*. Notre Dame: Notre Dame University Press, 1961.

D'Antonio, William V., and William H. Form. *Influentials in Two Border Cities*. Notre Dame: Notre Dame University Press, 1965.

Danzger, M. Herbert. "Community Power Structure: Problems and Continuities." *American Sociological Review* 29 (1964): 707–17.

Davis, James A. "Clustering and Balance Theory in Graphs." *Human Relations* 20 (1967): 181–87.

_____. "Clustering and Hierarchy in Interpersonal Relations: Testing Two Graph Theoretical Models on 742 Sociograms." *American Sociological Review* 35 (1970): 843–52.

Davis, James A., and Samuel Leinhardt. "The Structure of Positive Interpersonal Relations in Small Groups." In Joseph Berger, Morris Zelditch, Jr., and Bo Anderson (eds.), *Sociological Theories in Progress*, 2. Boston: Houghton Mifflin, 1971.

Davis, John P. *Corporations*. New York: Capricorn, 1961.

De George, Richard T., and Fernande M. De George. *The Structuralists: From Marx to Levi-Strauss*. Garden City, N.Y.: Doubleday, 1972.

Delaney, John L. "Network Dynamics for the Weak Tie Problem: A Simulation Study." Cambridge, Mass.: Harvard-Yale Preprints in Mathematical Sociology, 1978.

Dijkstra, Wil. "Response Bias in the Survey Interview: An Approach from Balance Theory." *Social Networks* 2 (1979/80): 285–304.

Dooley, Peter C. "The Interlocking Directorate." *American Economic Review* 59 (1969): 314–23.

Doreian, Patrick. "A Note on the Detection of Cliques in Valued Graphs." *Sociometry* 32 (1969): 237–42.

———. *Mathematics and the Study of Social Relations*. London: Weidenfeld and Nicolson, 1970.

———. "On the Evolution of Group and Network Structure." *Social Networks* 2 (1979/80): 235–52.

Durkheim, Emile. *Moral Education*. New York: Free Press, 1961.

———. *The Division of Labour in Society*. New York: Free Press, 1964a.

———. *The Rules of Sociological Method*. London: Free Press, 1964b.

Ehrmann, Jacques. *Structuralism*. Garden City, New York: Doubleday, 1970.

Engels, Frederich. *Anti-Dühring*. Moscow: Progress Publishers, 1969.

Epstein, A. L. "The Network and Urban Social Organization." *Rhodes-Livingston Journal* 29 (1961): 29–62.

Erickson, B. H. "Some Problems of Inference from Chain Data." In Karl F. Schuessler (ed.), *Sociological Methodology, 1979*. San Francisco: Jossey-Bass, 1979.

Evan, W. M. "An Organization-Set Model of Interorganizational Relations." In M. F. Tuite, M. Radnor, and R. R. Chisholm (eds.), *Interorganizational Decision Making*. Chicago: Aldine, 1972.

Fararo, Thomas J. *Mathematical Sociology*. New York: Wiley, 1973.

Fararo, Thomas J., and Morris H. Sunshine. *A Study of a Biased Friendship Net*. Syracuse: Syracuse University Press, 1964.

Faris, Robert E. L. *Chicago Sociology: 1920–1932*. San Francisco: Chandler, 1967.

Fennema, Meindart, and Huibert Schijf. "Analyzing Interlocking Directorates: Theory and Methods." *Social Networks* 1 (1978/79): 297–332.

Financial Post. Survey of Industrials. Toronto: Maclean Hunter, annual.

Fischer, Claude S.; Robert Max Jackson; C. Ann Stueve; Kathleen Gerson; and Lynne McCallister Jones; with Mark Baldassare. *Networks and Places: Social Relations in the Urban Setting*. New York: Free Press, 1977.

Fitch, Robert, and Mary Oppenheimer. "Who Rules the Corporations?" *Socialist Revolution* (1970): 73–107 (July/Aug.); 61–114 (Sept./Oct.); 33–94 (Nov./Dec.).

Flament, C. *Théorie des Graphes et Structure Sociale*. Paris: Mouton, 1965.

Foster, Brian L. "Formal Network Studies and the Anthropological Perspective." *Social Networks* 1 (1978/79): 241–55.

Foster, Caxton C.; Anatol Rapoport; and Carol J. Orwant. "A Study of a Large Sociogram, II: Elimination of Free Parameters." *Behavioral Science* 8 (1963): 56–65.

Foucault, Michel. *Madness and Civilization: A History of Insanity in the Age of Reason*. New York: Pantheon, 1965.

Frank, Ove. *Statistical Inference in Graphs*. Stockholm: Research Institute of National Defense, 1971.

_____. "Sampling and Estimation in Large Social Networks." *Social Networks* 1 (1978): 91–101.

Frank, Ove, and Frank Harary. "Balance in Stochastic Signed Graphs." *Social Networks* 2 (1980): 155–63.

Freeman, Linton C. *Patterns of Local Community Leadership*. Indianapolis, Ind.: Bobbs-Merrill, 1968.

_____. *A Bibliography of Social Networks*. Monticello, Ill.: Council of Planning Librarians, 1976.

_____. "A Set of Measures of Centrality Based on Betweenness." *Sociometry* 40 (1977): 35–41.

_____. "Segregation in Social Networks." *Sociological Methods and Research* 6 (1978): 411–29.

_____. "Centrality in Social Networks: Conceptual Clarification." *Social Networks* 1 (1978/79): 215–39.

Freeman, Linton C.; Warner Bloomberg, Jr.; Stephen P. Koff; Morris H. Sunshine; and Thomas J. Fararo. *Local Community Leadership*. Syracuse, N.Y.: University College of Syracuse University, 1960.

Freeman, Linton C.; Thomas J. Fararo; Warner Bloomberg, Jr.; and Morris H. Sunshine. *Metropolitan Decision-Making*. Syracuse, N.Y.: University College of Syracuse University, 1962.

_____. "Locating Leaders in Local Communities: A Comparison of Some Alternative Approaches." *American Sociological Review* 28 (1963): 791–98.

Freud, Sigmund. *The Problem of Anxiety*. New York: W. W. Norton, 1936.

_____. *The Ego and the Id*. Translated by Joan Riviere. London: Hogarth Press, 1950.

Friedman, Milton. *Price Theory*. Chicago: Aldine, 1976.

Friedmann, Harriet. "The Transformation of Wheat Production in the Era of the World Market, 1873–1935: A Global Analysis of Production and Exchange." Ph.D. dissertation, Harvard University, Cambridge, Mass., 1976.

_____. "Simple Commodity Production and Wage Labour in the American Plains." *Journal of Peasant Studies* 6 (1978a): 71–100.

_____. "World Market, State, and Family Farm: Social Bases of Household Production in the Era of Wage Labor." *Comparative Studies in Society and History* 20 (1978b): 545–86.

Gans, Herbert. *The Urban Villagers: Group and Class in the Life of Italian Americans*. New York: Free Press, 1962.

Gerth, Hans, and C. Wright Mills. *From Max Weber: Essays in Sociology*. New York: Oxford, 1946.

Gibbs, Jack. *Crime, Punishment, and Deterence.* New York: Elsevier, 1975.

Glanzer, M., and R. Glaser. "Techniques for the Study of Group Structure and Behavior: I. Analysis of Structure." *Psychological Bulletin* 56 (1959): 317–32.

Gordon, R. A. *Business Leadership in the Large Corporation.* Washington: Brookings Institution, 1943.

Gould, Peter, and Anthony Gatrell. "A Structural Analysis of a Game: The Liverpool v. Manchester United Cup Final of 1977." *Social Networks* 2 (1979/80): 253–73.

Granovetter, Mark. "The Strength of Weak Ties." *American Journal of Sociology* 78 (1973): 1360–80.

———. *Getting a Job.* Cambridge, Mass.: Harvard University Press, 1974.

———. "Network Sampling: Some First Steps." *American Journal of Sociology* 81 (1976): 1287–1303.

Gray, W., and N. Rizzo (eds.). *Unity through Diversity: Festschriff in Honor of Ludwig von Bertalanffy.* New York: Gordon & Breach, 1971.

Gurevitch, M. "The Social Structure of Acquaintanceship Networks." Ph.D. dissertation, Massachusetts Institute of Technology, Cambridge, Mass., 1961.

Gurevitch, M., and A. Weingrod. "Who Knows Whom: Contact Networks in the Israeli National Elite. *Megamot* 22 (1976): 357–78.

Guttman, Louis. "A General Nonmetric Technique for Finding the Smallest Coordinate Space for a Configuration of Points." *Psychometrika* 33 (1968): 469–506.

Hagen, Everett E. *The Economics of Development.* Homewood, Ill.: Irwin, 1968.

Hallinan, Maureen T. *The Structure of Positive Sentiment.* New York: Elsevier, 1974.

———. "The Process of Friendship Formation." *Social Networks* 1 (1978/79): 193–210.

Hamburg, Morris. *Statistical Analysis for Decision Making.* 2nd ed. New York: Harcourt, Brace, Jovanovich, 1977.

Hannerz, U. "Networks and Culture in a Black American Ghetto." *Ethnos* 32 (1967): 1–9, 35–60.

Harary, Frank. "Graph Theoretic Measures in the Management Sciences." *Management Science* 5 (1959): 387–403.

Harary, Frank; Robert Z. Norman; and Dorwin Cartwright. *Structural Models: An Introduction to the Theory of Directed Graphs.* New York: John Wiley & Sons, 1965.

Harries-Jones, Peter. " 'Home-boy' Ties and Political Organization in a Copperbelt Township." In J. Clyde Mitchell (ed.), *Social Networks in Urban Situations: Analyses of Personal Relationships in African Towns.* Manchester: Manchester University Press, 1969.

Harsanyi, John C. *Rational Behavior and Bargaining Equilibrium in Games and Social Situations.* New York and Cambridge: Cambridge University Press, 1977.

Hart, H. L. A. *The Concept of Law*. Oxford: Clarendon Press, 1961.

Hawkes, Terence. *Structuralism and Semiotics*. Berkeley and Los Angeles, Calif.: University of California Press, 1977.

Heider, Fritz. "Attitudes and Cognitive Organization." *Journal of Psychology* 21 (1946): 107–12.

Heil, Greg H., and Harrison C. White. "An Algorithm for Finding Simultaneous Homomorphic Correspondences between Graphs and Their Image Graphs." *Behavioral Science* 21 (1976): 26–35.

Heiss, C. A. *Accounting in the Administration of Large Business Enterprises*. Cambridge, Mass.: Harvard University Press, 1943.

Hilferding, Rudolph. *Das Finanzkapital*. 1910. Reprint ed. Frankfurt am Main: Europäische Verlaganstalt, 1968.

Hobbes, Thomas. *De Cive* or *The Citizen*. New York: Appleton-Century-Crofts, 1949.

Holland, Paul, and Samuel Leinhardt. "A Unified Treatment of Some Structural Models for Sociometric Data." Pittsburgh: Carnegie Mellon University, Technical Report, 1970a.

_____. "A Method for Detecting Structure in Sociometric Data." *American Journal of Sociology* 70 (1970b): 492–513.

Howard, B. B., and M. Upton. *Introduction to Business Finance*. New York: McGraw-Hill, 1953.

Howard, Leslie. "Industrialization and Community in Chotangpur." Ph.D. dissertation, Harvard University, Cambridge, Mass., 1974.

Howell, Nancy (Lee). *The Search for an Abortionist*. Chicago: University of Chicago Press, 1969.

_____. *Demography of the Dobe !Kung*. New York: Academic Press, 1979.

Howell, Nancy, and Victor Lehotay. "AMBUSH: A Computer Program for Stochastic Microsimulation of Small Human Populations." *American Anthropologist* 80 (1978): 905–22.

Hubbell, C. "An Input-Output Approach to Clique Identification." *Sociometry* 28 (1965): 377–99.

Hunter, Floyd. *Community Power Structure: A Study of Decision Makers*. Garden City, New York: Doubleday, 1953.

Jacobson, D. "Network Analysis in East Africa: The Social Organization of Urban Transients." *Canadian Review of Sociology and Anthropology* 7 (1970): 281–86.

Jahoda, Marie. *Current Concepts of Positive Mental Health*. New York: Basic, 1958.

Jakobson, Roman. *Selected Writings*. 4 vols. The Hague: Mouton, 1962.

Jakobson, Roman, and Morris Halle. *Fundamentals of Language*. The Hague: Mouton, 1956.

Jedlicka, Davor. "Opportunities, Information Networks, and International Migration Streams." *Social Networks* 1 (1978/79): 277–84.

Jennings, Helen. "Structure of Leadership: Development and Sphere of Influence." *Sociometry* 1 (1937): 99–143.

Kadushin, Charles. "The Friends and Supporters of Psychotherapy: On Social Circles in Urban Life." *American Sociological Review* 31 (1966): 786–802.

———. "Power, Influence and Social Circles: A New Methodology for Studying Opinion Makers." *American Sociological Review* 33 (1968): 685–99.

Kapferer, Bruce. "Norms and the Manipulation of Relationships in a Work Context." In J. Clyde Mitchell (ed.), *Social Networks in Urban Situations: Analyses of Personal Relationships in African Towns.* Manchester: Manchester University Press, 1969.

Kaplan, Abraham. *The Conduct of Inquiry: Methodology for Behavioral Science.* Scranton, Pa.: Chandler, 1964.

Katz, Pearl. "Acculturation and Social Networks of American Immigrants in Israel." Ph.D. dissertation, SUNY Buffalo, 1974.

Kemeny, John G., and J. Laurie Snell. *Mathematical Models in the Social Sciences.* Boston: Ginn, 1962.

Key, V. O. *Politics, Parties and Pressure Groups.* New York: Knopf, 1959.

Killworth, Peter, and H. Russell Bernard. "Catij: A New Sociometric and its Application to a Prison Living Unit." *Human Organization* 33 (1974): 335–50.

———. "Informant Accuracy in Social Network Data." *Human Organization* 35 (1976): 269–96.

———. "The Reverse Small-World Experiment." *Social Networks* 1 (1978/79): 159–93.

Kim, Ki Hang, and Fred William Roush. *Mathematics for Social Scientists.* New York: Elsevier, 1980.

Klir, George J. (ed.). *Trends in General Systems Theory.* New York: Wiley, 1972.

Knowles, James. "The Rockefeller Financial Group." Warner Modular Publications, No. 343, 1973.

Kornhauser, William. *The Politics of Mass Society.* Glencoe, Ill.: Free Press, 1959.

Korte, C., and S. Milgram. "Acquaintanceship Networks between Racial Groups: Application of the Small World Method." *Journal of Personality and Social Psychology* 15 (1970): 101–8.

Kruskal, J. B. "Nonmetric Multidimensional Scaling: A Numerical Method." *Psychometrika* 29 (1964): 115–29.

Kuhn, Thomas. *The Structure of Scientific Revolutions.* Chicago: University of Chicago Press, 1970.

Lane, Michael. *Structuralism: A Reader.* London: Cape, 1970.

Laslo, E. *Introduction to Systems Philosophy.* New York: Gordon & Breach, 1971.

Laumann, Edward O. *Prestige and Association in an Urban Community.* Indianapolis: Bobbs-Merrill, 1966.

———. *Bonds of Pluralism: The Form and Substance of Urban Social Networks.* New York: Wiley, 1973.

Laumann, Edward O., and Louis Guttman. "The Relative Associational Contiguity of Occupations in an Urban Setting." *American Sociological Review* 31 (1966): 169–78.

Laumann, Edward O., and Peter V. Marsden. "The Analysis of Oppositional Structures in Political Elites: Identifying Collective Actors." *Sozialwissenschaftliche Schriften* 14 (1978): 127–80; *American Sociological Review* 44 (1979): 713–32 (later version).

Laumann, Edward O.; Peter V. Marsden; and Joseph Galaskiewicz. "Community-Elite Influence Structures: Extension of a Network Approach." *American Journal of Sociology* 83 (1977): 594–631.

Laumann, Edward O., and Franz Pappi. "New Directions in the Study of Community Elites." *American Sociological Review* 38 (1973): 212–30.

_____. *Networks of Collective Action: A Perspective on Community Influence Systems.* New York: Academic Press, 1976.

Lazarsfeld, Paul F. (ed). *Mathematical Thinking in the Social Sciences.* New York: Russell & Russell, 1954.

Leach, Edmund. *Claude Lévi-Strauss.* New York: Viking Press, 1970.

Leeds, Anthony. "The Culture of Poverty: Conceptual, Logical, and Empirical Problems with Perspectives from Brazil and Peru." In E. Leacock (ed.), *The Culture of Poverty: A Critique.* New York: Simon and Schuster, 1970.

Leinhardt, Samuel (ed.). *Social Networks: A Developing Paradigm.* New York: Academic Press, 1977.

Levine, Joel H. "The Sphere of Influence." *American Sociological Review* 37 (1972): 14–27.

Levinson, Charles. *Capital, Inflation, and the Multinationals.* Winchester, Mass.: Allen & Unwin, 1971.

Lévi-Strauss, Claude. *The Elementary Structures of Kinship.* Boston: Beacon Press, 1969.

Lin, Nan. "Stratification of the Formal Communication System in American Sociology." *American Sociologist* 9 (1974): 199–206.

Lin, Nan, and Paul W. Dayton. "The Urban Communication Network and Social Stratification: A Small World Experiment." Paper presented at the annual meetings of the International Communication Association, Portland, Oregon, 1976.

Lin, Nan, and C. E. Nelson. "Bibliographic Reference Patterns in Core Sociology Journals." *American Sociologist* 4 (1969): 47–50.

Lintner, John. "The Financing of Corporations." In Edward Mason (ed.), *The Corporation in Modern Society.* New York: Atheneum, 1967.

Lorrain, François. *Réseaux Sociaux et Classifications Sociales.* Paris: Hermann, 1975.

Lorrain, François, and Harrison C. White. "Structural Equivalence of Individuals in Social Networks." *Journal of Mathematical Sociology* 1 (1971): 49–80.

Luce, R. D., and A. W. Tucker (eds.). *Contributions to the Theory of Games, IV.* Princeton: Princeton University Press, 1959.

Lundberg, C. "Patterns of Acquaintanceship in Society and Complex Organization: A Comparative Study of the Small World Problem." *Pacific Sociological Review* 18 (1975): 206–22.

Lynd, Robert S., and Helen M. Lynd. *Middletown*. New York: Harcourt, Brace, 1929.

Marsden, Peter V., and Edward O. Laumann. "Collective Action in a Community Elite: Exchange, Influence Resources, and Issue Resolution." In R. J. Liebert and A. W. Imershein (eds.), *Power, Paradigms, and Community Research*. Beverly Hills: Sage, 1977.

Marx, Karl. *Capital*. Moscow: Progress Publishers, n.d.

———. *A Contribution to the Critique of Political Economy*. Moscow: Progress Publishers, n.d.

———. *Selected Works*. New York: International Publishers, n.d.

———. *Economic and Philosophic Manuscripts of 1844*. Moscow: Foreign Language Pub., 1961.

Maxwell, Lee M., and Myril B. Reed. *The Theory of Graphs: A Basis for Network Theory*. New York: Pergamon, 1971.

Mayer, Adrian. "The Significance of Quasi-Groups in the Study of Complex Societies." In M. Banton (ed.), *The Social Anthropology of Complex Societies*. London: Tavistock, 1966.

Mayhew, Leon. *Society, Institutions, and Activity*. Glenview, Ill.: Scott, Foresman, 1971.

McFarland, David D., and Daniel J. Brown. "Social Distance as a Metric: A Systematic Introduction to Smallest Space Analysis." In Edward O. Laumann (ed.), *Bonds of Pluralism: The Form and Substance of Urban Social Networks*. New York: Wiley, 1973.

Mead, George H. *Mind, Self, and Society*. Chicago: University of Chicago Press, 1934.

Menken, J. A. "Biological Determinants of Demographic Processes." *American Journal of Public Health* 64 (1974): 657–61.

Menzel, Herbert, and Elihu Katz. "Social Relations and Innovation in the Medical Profession: The Epidemiology of a New Drug." *Public Opinion Quarterly* 19 (1955): 337–52.

Merton, Robert K. *Social Theory and Social Structure*. New York: Free Press, 1968.

Milgram, Stanley. "The Small World Problem." *Psychology Today* 1 (1967): 61–67.

———. "Interdisciplinary Thinking and the Small World Problem." In M. Sherif and C. Sherif (eds.), *Interdisciplinary Relationships in the Social Sciences*. Chicago: Aldine, 1969.

Miller, Delbert. "Democracy and Decision-Making in the Community Power Structure." In William V. D'Antonio and Howard J. Ehrlich (eds.), *Power and Democracy in America*. Notre Dame: Notre Dame University Press, 1961.

Mills, C. Wright. *White Collar: The American Middle Classes*. New York: Oxford, 1951.

_____. *The Power Elite.* New York: Oxford, 1956.

Mintz, Beth; Peter Freitag; Carol Hendricks; and Michael Schwartz. "Problems of Proof in Elite Research." *Social Problems* 23 (1976): 315–24.

Mintz, Beth, and Michael Schwartz. "The Role of Financial Institutions in Interlock Networks." Paper read at the New Directions in Structural Analysis Conference, Toronto, 1978; *American Sociological Review,* forthcoming.

Mitchell, J. Clyde. "The Study of African Urban Structures." In *Inter-African Conference on Housing and Urbanization, Second Session, Nairobi, 1959.* London: Scientific Council for Africa South of the Sahara, 1959.

_____ (ed.). *Social Networks in Urban Situations: Analyses of Personal Relationships in Central African Towns.* Manchester: Manchester University Press, 1969.

Moody's Banking and Finance Manual. New York: Moody's Investors' Service (annual).

Moody's Industrial Manual. New York: Moody's Investors' Service (annual).

Moreno, J. L. *Sociometry, Experimental Method and Science of Society.* Beacon, New York: Beacon House, 1951.

Mullins, Nicholas. *Theory and Theory Groups in Contemporary American Sociology.* New York: Harper & Row, 1973.

Nadel, S. F. *The Theory of Social Structure.* Glencoe, Ill.: Free Press, 1957.

Niemoller, Kees, and Bert Schijf. "Applied Network Analysis." *Quality and Quantity* 14 (1980): 101–16.

Norman, R. Z., and F. S. Roberts. "A Measure of Relative Balance for Social Structures." In J. Berger, M. Zelditch, and Bo Anderson (eds.), *Sociological Theories in Progress, 2.* New York: Houghton Mifflin, 1972.

Oglesby, Carl. *The Yankee and Cowboy War.* New York: Berkley, 1977.

Olinick, Michael. *An Introduction to Mathematical Models in the Social and Life Sciences.* Reading, Mass.: Addison-Wesley, 1978.

Parsons, Talcott. "The Law and Social Control." In William E. Evan (ed.), *Law and Society.* New York: Free Press, 1962.

Perrucci, Robert, and Marc Pilisuk. "Leaders and Ruling Elites: The Interorganizational Bases of Community Power." *American Sociological Review* 35 (1970): 1040–57.

Piaget, Jean. *Structuralism.* London: Routledge & Kegan Paul, 1971.

Pool, Ithiel de Sola, and Manfred Kochen. "Contacts and Influence." *Social Networks* 1 (1978/79): 5–51.

Poucke, Willy van. "Network Constraints on Social Action: Preliminaries for a Network Theory." *Social Networks* 2 (1979/80): 181–90.

Rapoport, Anatol. "Nets with Distance Bias." *Bulletin of Mathematical Biophysics* 13 (1951): 85–91.

_____. "Ignition Phenomena in Random Nets." *Bulletin of Mathematical Biophysics* 14 (1952): 35–44.

———. "The Diffusion Problem in Mass Behavior." *General Systems* 1 (1956): 48–55.

———. "Contribution to the Theory of Random and Biased Nets." *Bulletin of Mathematical Biophysics* 19 (1957): 257–77.

———. "Nets with Reciprocity Bias." *Bulletin of Mathematical Biophysics* 20 (1958): 191–201.

———. "Mathematical Aspects of General Systems Theory." *General Systems* 11 (1966): 3–11.

———. *N-Person Game Theory: Concepts and Applications.* Ann Arbor: University of Michigan Press, 1970.

———. "The Uses of Mathematical Isomorphism in General Systems Theory." In George J. Klir (ed.), *Trends in General Systems Theory.* New York: Wiley, 1972.

———. *Two Person Game Theory: The Essential Ideas.* Ann Arbor: University of Michigan Press, 1973.

———. *Fights, Games, and Debates.* Ann Arbor: University of Michigan Press, 1974.

———. "A Probabilistic Approach to Networks." *Social Networks* 2 (1979/80): 1–18.

Rapoport, Anatol, and William J. Horvath. "A Study of a Large Sociogram." *Behavioral Science* 6 (1961): 279–91.

Reid, Samuel Richardson. *The New Industrial Order: Concentration, Regulation, and Public Policy.* New York: McGraw-Hill, 1976.

Rogers, D. L. "Sociometric Analysis of Interorganizational Relations: Application of Theory and Measurement." *Rural Sociology* 39 (1974): 487–503.

Rogers, Everett M. *Diffusion of Innovations.* New York: Free Press, 1962.

Rohner, Ronald. *The Ethnography of Franz Boas.* Chicago: University of Chicago Press, 1969.

Rose, Arnold. *The Power Structure: Political Process in American Society.* New York and London: Oxford University Press, 1967.

Rossi, Peter H. "Power and Community Structure." *Midwest Journal of Political Science* 4 (1960): 390–401.

Rousseau, Jean-Jacques. *The Social Contract* and *Discourses.* London: Dent, 1973.

Royden, H. L. *Real Analysis.* New York: Macmillan, 1963.

Runger, George, and Stanley Wasserman. "Longitudinal Analysis of Friendship Networks." *Social Networks* 2 (1979/80): 143–54.

Safilios-Rothschild, Constantina. "Towards a Cross-Cultural Conception of Family Modernity." *Journal of Comparative Family Studies* 1 (1970): 17–25.

Sailer, Lee D. "Structural Equivalence: Meaning and Definition, Computation and Application." *Social Networks* 1 (1978): 73–90.

St.-Simon, Claude Henri Comte de. *Social Organization, the Science of Man.* Edited and translated by Felix Markham. New York: Harper & Row, 1964.

Salaff, Janet. *Working Daughters of Hong Kong: Filial Piety or Power in the Family?* Cambridge and New York: Cambridge University Press, 1981.

Sampson, Anthony. *Anatomy of Britain.* London: Holder and Stoughton, 1962.

Sapir, Edward. *Language.* New York: Harcourt, Brace, 1921.

Saussure, Ferdinand de. *Course in General Linguistics.* New York: McGraw-Hill, 1966.

Schwartz, J. E. "An Examination of CONCOR and Related Methods for Blocking Sociometric Data." In D. R. Heise (ed.), *Sociological Methodology, 1977.* San Francisco: Jossey-Bass, 1977.

Scott, John. *Corporations, Classes and Capitalism.* London: Hutchinson, 1979.

Seidman, Stephen B., and Brian L. Foster. "A Note on the Potential for Genuine Cross-Fertilization between Anthropology and Mathematics." *Social Networks* 1 (1978): 65–72.

Selznick, Phillip. "Sociology and Natural Law." *Natural Law Forum* 6 (1964): 84–108.

Shaw, M. E. "Group Structure and the Behavior of Individuals in Small Groups." *Journal of Psychology* 38 (1954): 139–49.

Shepard, R. N. "The Analysis of Proximities: Multidimensional Scaling with an Unknown Distance Function, I." *Psychometrika* 27 (1962): 125–40.

———. "The Analysis of Proximities: Multidimensional Scaling with an Unknown Distance Function, II." *Psychometrika* 27 (1962): 219–46.

———. "Metric Structures in Ordinal Data." *Journal of Mathematical Psychology* 3 (1966): 287–315.

Shepard, R. N., and Phipps Arabie. "Additive Clustering: Representation of Similarities as Combinations of Discrete Overlapping Properties." *Psychological Review* 86 (1979): 87–123.

Shepard, R. N., and J. D. Carroll. "Parametric Representation of Nonlinear Data Structures." In P. R. Krishnaiah (ed.), *International Symposium on Multivariate Analysis* (Dayton, Ohio, 1965). New York: Academic Press, 1966.

Shepard, R. N.; A. Kimball Romney; and Sara Beth Nerlove (eds.). *Multidimensional Scaling: Theory and Applications in the Behavioral Sciences.* New York: Seminar Press, 1972.

Shibutani, Tamatsu. *Society and Personality: An Interaction Approach to Social Psychology.* Englewood Cliffs, N.J.: Prentice-Hall, 1961.

Siddall, Roger B. *A Survey of Large Law Firms in the United States.* New York: Vantage, 1956.

Simmel, Georg. "The Metropolis and Mental Life." In Kurt H. Wolff (ed. and trans.), *The Sociology of Georg Simmel.* Glencoe, Ill.: Free Press, 1950.

———. "The Stranger." In Kurt H. Wolff (ed. and trans.), *The Sociology of Georg Simmel.* Glencoe, Ill.: Free Press, 1950.

Smigel, Erwin O. *The Wall Street Lawyer*. Bloomington: Indiana University Press, 1969.

Snyder, D., and E. L. Kick. "Structural Position in the World System and Economic Growth, 1955–70: A Multiple Network Analysis of Transnational Interactions." *American Journal of Sociology* 84 (1979): 1096–126.

Sonquist, John A., and Thomas Koenig. "Interlocking Directorates in the Top U.S. Corporations: A Graph Theory Approach." *Insurgent Sociologist* 5 (1976): 196–229.

Spence, A. Michael. *Market Signaling: Informational Transfer in Hiring and Related Screening Processes*. Cambridge, Mass.: Harvard University Press, 1974.

Standard & Poor's Register of Corporations, Directors, and Executives. New York: Standard & Poor's Corporation (annual).

Statistics Canada. *Industrial Organization and Concentration in the Manufacturing and Logging Industries*. Ottawa: Information Canada, 1973.

Stein, Maurice. *The Eclipse of Community*. Princeton, N.J.: Princeton University Press, 1960.

Tepperman, Lorne. "The Persistence of Dynasties: A Structural Perspective." In S. D. Berkowitz and Barry Wellman (eds.), *Structural Sociology*. Cambridge and New York: Cambridge University Press, forthcoming.

Thom, René. *Structural Stability and Morphogenesis*. Translated by D. H. Fowler. Reading, Mass.: W. A. Benjamin, 1975.

Tilly, Charles. "Food Supply and Public Order in Modern Europe." In Charles Tilly (ed.), *The Formation of National States in Western Europe*. Princeton: Princeton University Press, 1975.

――――. *From Mobilization to Revolution*. Reading, Mass.: Addison-Wesley, 1978.

Travers, J., and S. Milgram. "An Experimental Study of the Small World Problem." *Sociometry* 32 (1969): 425–43.

U.S. Federal Trade Commission. *Report of the Federal Trade Commission on Interlocking Directorates*. Washington: U.S. Govt. Printing Office, 1951.

U.S. National Resources Committee. *The Structure of the American Economy, Part I: Basic Characteristics*. Washington: U.S. Govt. Printing Office, 1939.

Useem, Michael. "Corporations and the Corporate Elite." *Annual Review of Sociology* 6 (1980): 41–77.

Utton, M. A. *Industrial Concentration*. Middlesex: Penguin, 1970.

Vico, Giambattista. *Principi Di Scienza Nuova*. 1725.

Von Neumann, John, and Oskar Morgenstern. *Theory of Games and Economic Behavior*. New York: Wiley, 1944.

Walker, Robert L. "Social and Spatial Constraints in the Development and Functioning of Social Networks: A Case Study of Guildford." Ph.D. dissertation, London School of Economics, 1974.

Walton, John. "Substance and Artifact: The Current Status of Research on Community Power Structure." *American Journal of Sociology* 71 (1966): 430–38.

Warner, W. L., and D. B. Unwalla. "The System of Interlocking Directorates." In W. L. Warner, D. B. Unwalla, and J. H. Trimm (eds.), *The Emergent American Society*. Vol. I. New Haven: Yale University Press, 1967.

Warner, W. L.; D. B. Unwalla; and J. H. Trimm (eds.). *The Emergent American Society*. Vol. I. New Haven: Yale University Press, 1967.

Wasserman, Stanley. "A Stochastic Model for Directed Graphs with Transition Rates Determined by Reciprocity." In Karl F. Schuessler (ed.), *Sociological Methodology, 1980*. San Francisco: Jossey-Bass, 1980.

Waverman, L., and R. Baldwin. "Determinants of Interlocking Directorates." Toronto: Institute for Policy Analysis, 1975.

Wellman, Barry. "Urban Connections." Toronto: Centre for Urban and Community Studies, University of Toronto, 1976.

_____. "The Community Question: The Intimate Networks of East Yorkers." *American Journal of Sociology* 84 (1979): 1201–31.

_____. "Network Analysis: From Metaphor and Method to Theory and Substance." *Sociological Theory* 1, forthcoming.

Wellman, Barry, and Marilyn Whitaker (eds.). *Community-Network-Communications: An Annotated Bibliography*. Bibliographic Paper no. 4. 2nd ed. Toronto: Centre for Urban and Community Studies, University of Toronto, 1974.

Wheeldon, Prudence. "The Operation of Voluntary Associations and Personal Networks in the Political Processes of an Inter-ethnic Community." In J. Clyde Mitchell (ed.), *Social Networks in Urban Situations: Analyses of Personal Relationships in African Towns*. Manchester: Manchester University Press, 1969.

White, Douglas. "Mathematical Anthropology." In J. Honigmann (ed.), *Handbook of Social and Cultural Anthropology*. Chicago: Rand-McNally, 1973.

_____. "Material Entailment Analysis: Theory and Illustrations." Research Report 15. School of Social Sciences, University of Calif., Irvine, 1980.

_____. "Structural Equivalences in Social Networks: Concepts and Measurement of Role Structures." Irvine, Calif.: School of Social Sciences, University of California, Irvine, MS, 1980. (Prepared for the Laguna Beach Conference on Research Methods in Social Network Analysis.)

White, Harrison C. *An Anatomy of Kinship*. Englewood Cliffs., N.J.: Prentice-Hall, 1963.

_____. "Search Parameters for the Small World Problem." *Social Forces* 49 (1970a): 259–64.

_____. *Chains of Opportunity: System Models of Mobility in Organizations*. Cambridge, Mass.: Harvard University Press, 1970b.

_____. "Everyday Life in Stochastic Networks." *Sociological Inquiry* 43 (1973): 43–49.

_____. "Subcontracting with an Oligopoly: Spence Revisited." *RIAS Program Working Paper #1*. Cambridge, Mass.: Harvard University, 1976.

_____. "On Markets." *RIAS Program Working Paper #16*. Cambridge, Mass.: Harvard University, 1979.

_____. "Markets as Stages for Producers." Toronto: University of Toronto, Structural Analysis Programme, 1980.

_____. "Production Markets as Induced Role Structures." In Samuel Leinhardt (ed.), *Sociological Methodology, 1981*. San Francisco: Jossey-Bass, 1981.

_____. "Varieties of Markets." In S. D. Berkowitz and Barry Wellman (eds.), *Structural Sociology*. Cambridge and New York: Cambridge University Press, forthcoming.

_____. "Where Do Markets Come From?" *American Journal of Sociology*, forthcoming.

White, Harrison C.; Scott A. Boorman; and Ronald L. Breiger. "Social Structure from Multiple Networks, I: Blockmodels of Roles and Positions." *American Journal of Sociology* 81 (1976): 730–80.

Wiener, Norbert. *Cybernetics*. New York: Wiley, 1948.

Wolff, Kurt H. (ed. and trans.). *The Sociology of Georg Simmel*. Glencoe, Ill.: Free Press, 1950.

Young, Kimball. *Personality and Problems of Adjustment*. New York: Appleton-Century-Crofts, 1947.

Young, Michael, and Peter Willmott. *Family and Kinship in East London*. London, Penguin, 1962.

_____. *The Symmetrical Family*. New York: Pantheon, 1973.

Zeitlin, Maurice; Lynda Ann Ewen; and Richard Earl Ratcliff. " 'New Princes' for Old? The Large Corporation and the Capitalist Class in Chile." *American Journal of Sociology* 80 (1975): 87–123.

Author Index

Subject Index

acquaintanceship
   volumes of, 64
actors
   collective, definition of, 129–31
   positional, 61
   reciprocal orientation of, 26
adjacency, 63, 120. *See* centrality.
   as limiting case of general property, 15
   as measured by directorship interlocks, 82
   as used in centrality measures, 18
   corporate ties and, 82
   definition of, 15
   path-distance and, 18, 40
   second-order, 19
aggregative models, *see* models, aggregative
algebraic models, *see* models, algebraic
"Algebra of Kinship, The," 4. *See* Boyd, John Paul.
amorphous model, *see* models, of elites
analogs, 168 n.33
   in structural modeling, 153
   modeling and, 14
"Analysis of Oppositional Structures in Political Elites, The," 129. *See* Laumann, Edward O. and Marsden, Peter V.
Anatomy of Kinship, An, 54. *See* White, Harrison C.
anthropology
   early network studies in, 3
antipluralists, 126, 191 n.5
ascriptive groups, *see* groups, ascriptive

backcloth, 158
balance
   cognitive theory and, 20
   in graphs, definition of, 20
behavior, 62, 154
   of given corporations, 72, 103–15
   relationship of morphology and structure to, 2
bias, *see* networks
binary matrices, *see* matrices, binary
blockings, *see* blockmodeling

blockmodel(s)(ing), 9, 133, 135–37, 156, 157. *See also* models, algebraic.
   blocs in, 133–37, 169 n.41, 194 n.36, 37, 38
   definition of, 169
   image graph in, 133
   of social structures, 133
blocs, *see* blockmodeling
b-measure
   definition of, 200 n.42
Bonds of Pluralism: The Form and Substance of Urban Social Networks, 57, 58, 61–63. *See* Laumann, Edward O.
boundaries, 113–15, 119
bounded groups, 3, 4

capital, 117
   markets, *see* markets
   mobilization, 117
   pools, *see* enterprises
   relationship of structure to that of upper class group, 117
categories
   as basis of conventional analysis, 2, 14
   received, as used in pictorial realism and nominalism, 4
centrality
   measures of
      adjacency notion of, 19. *See* adjacency.
      geodesic, 19
      inverse, 19
      Stony Brook group's studies of, 111
chain(s)
   contact, 65
   length, 65
   members, 65
choice-constraint models, *see* models, choice-constraint
classes
   as determined by structural analysts, 14
cliques
   definition of, 15, 171 n.54
   relationship of clusters to, 15
cliquishness
   definition of, 28, 176 n.13
closed systems, *see* systems